AF715690

INFORMATION TECHNOLOGY AND AFRICA
Riding the Fourth Industrial Revolution

INFORMATION

TECHNOLOGY

AND

AFRICA

Riding the Fourth

Industrial Revolution

by

Evans E. Woherem

PRIMA

Copyright © 2023 by Evans Woherem

The right of Evans Woherem to be identified as the
author of this work has been asserted by him in
accordance with the copyright laws.

All rights reserved.

No part of this publication may be reproduced, transmitted,
or stored in a retrieval system, in any form or by any means,
without permission in writing from the copyright holder.

ISBN: 978-978-59760-5-2 (Paperback)
978-978-59760-8-3 (Hardcover)

Published in Nigeria in 2023 by Narrative Landscape Press
under its Prima imprint
1B Olatunde Ayoola Avenue, Obanikoro, Lagos State, Nigeria
07014522083, 09090554406 – 7
contact@narrativelandscape.com
www.narrativelandscape.com

A catalogue record for this book is available from
the National Library of Nigeria.

Cover Design and Layout by Adebayo Gbenga

DEDICATION

This book is primarily dedicated to the youths of Africa as they work frantically toward tech-driven economic development in Africa.

Secondly, it is dedicated to the late Professor Calestous Juma, who inspired me greatly on the need for a technology-led development of Africa.

Finally, the book is dedicated to my children, Chizi and Chioma, to guide them toward a new Africa.

CONTENTS

FOREWORD

In the early pages of this book, Dr Evans Woherem describes the *Sankofa* bird. It's an African word from the Akan tribe in Ghana. The literal translation of the word and the symbol is *that you must be able to look to the past and bring forward what is most beneficial to successfully forward to the future.* Sankofa epitomises Dr Woherem and his vision for Africa's unique journey towards its future as a significant player in the Fourth Industrial Revolution while always being mindful of Africa's glorious past.

I first met Dr Evans Woherem in the summer of 2014 in Lagos, Nigeria, when he was kind enough to invite me to share my American perspective of a technological society at an early Digital Africa conference. At that conference, despite being held in one of Lagos's nicest hotels, power blackouts and Internet outages were routine parts of each day.

It struck me that the things I took for granted in America, like uninterrupted high-speed Internet, could not be taken for granted in Nigeria. And yet, this notion of not taking things for granted was precisely the thing that spawned a resilience and determination that would be the major catalyst for pushing Africa ahead in its digital mission. At that early conference, led by Dr Woherem, I saw a clarity of purpose that inspired and challenged attendees to overcome obstacles and get on with the work of creating a tech-forward continent.

I also remember thinking to myself, "Why was 'this guy' so intent on his mission to lead Africa into the Fourth Industrial Revolution?" He certainly didn't need the income. Woherem was already a successful Harvard Business School alumni with a stellar career in management and organisational structure. As I got to know him over the years, it became clear that guiding Africa towards its rightful place in a digital world was more than a job for him. It was a passion and a life's mission. A natural leader, Woherem is driven by sharing his learnings to create an inclusive way to raise Africa to its place as a leader of modern technological innovation.

The final thing that struck me about Dr Woherem was that he was a true polymath. Someone who could speak eloquently about everything from Africa's majestical past, to how to train and acquire manpower, to social-cultural norms, to the power of AI and cloud computing. His gift is the ability to weave disparate pieces of knowledge from all disciplines into a compelling narrative. The polymath approach to IT and Africa is the book's recipe for success.

What Dr Woherem has created is a blueprint for how Africa can move forward. But it's also a book for anyone from any country seeking a deeper understanding of how the technologies propelling us into the future can be systematically harnessed and implemented.

For as long as I've known Dr Evans Woherem, he has set his sights on the Fourth Industrial Revolution as Africa's ticket to the modern global economy. Africa, often called the cradle of civilisation, has "the right stuff" to continue evolving. Through this instructive and far-reaching book, Woherem makes it abundantly clear that we look to the past to inform the future. He is someone who has personally taught me to look at the gestalt of our experience and use that to tell where we go next. I know that he'll do the same for you, his readers.

Robin Raskin
Founder, The Virtual Events Group (VEG)

PREFACE

For quite a while, I have been an advocate of the adoption, diffusion, and active development of digital technology in Africa for the continent's economic growth. Early on, I realised that through technology, African countries could emancipate Africa.

As I worked on my first master's degree, and through the time I began my employment with Lloyd's Register, UK, I dove deep into building extremely viable computer systems to create technical marvels of economic value, user-friendliness, social and business viability, productivity, and ultimately, sustainability. I subsequently developed a methodology for capturing such systems' best possible technical and socio-organisational requirements.

Along the way, my path crossed with that of Professor Calestous Juma, who, at the time, was with the Science Policy Research Unit of the University of Sussex, UK. Through my interactions with Professor Juma, my passion began to expand beyond building computer systems with unique functional and economic requirements to how African nations could leverage technology to leapfrog the continent's economic development.

The resultant effect of this was twofold. Firstly, in my second master's degree, I focused my dissertation on the topic of "Microelectronics and Africa." Secondly, I wrote a book titled *Information Technology in Africa: Challenges and Opportunities*. My motivation for these efforts was to instigate a drive towards the development of Africa via technology – more specifically, Information Technology.

However, since I wrote the book in 1993, Information Technology has advanced in leaps and bounds, and a new book about coming to terms with its three-decade metamorphosis and evaluating the state of the economic growth of African nations over the same period is long overdue.

This new book explores a gap between Africa and the rest of the world, and it compels its readers on the need for Africa to act decisively to close this gap as soon as possible. Technology is defining the world in the 21st century. If African nations fail to catch up with the developed world technologically, it will be impossible for them to compete in this global economy.

In the continued efforts to advocate for the adoption and vigorous development of information technology in Africa, I spearheaded the establishment of a digital technology think-tank and advocacy group called Digital Africa, an initiative geared towards bridging the gap between Africa and the rest of the world concerning Information Technology. We realise that this gap concerning Information Technology is enormous. Given the importance of technology as we progress through the 21st century, Africa must play a more prominent role in the use of technology and technological development.

Since its inception in 2011, we have faithfully organised an annual Digital Africa Exhibition and Conference, a platform to prepare the continent for emerging technologies preparative to the Fourth Industrial Revolution. This revolution is evolving exponentially rather than linearly and is disrupting almost every industry globally.

Digital Africa has also served as a platform for displaying and demonstrating emerging technologies trending across the globe. At the end of the yearly conference, we develop a report, a communique, some new goals, and action points that the relevant stakeholders and key players in the Information Technology sectors of African countries should implement.

This book is a continuation of all the efforts I have been making and the advocacy I have been committed to all these years. I intend to highlight the need for African countries to make technology and innovation a strategic priority to integrate them into the global knowledge economy fully. Such integration will liberate them from the oppressive dominance of globalisation.

The book is also a rallying cry for African countries and leaders to take these issues seriously. It aims to inspire Africans by making them realise that achieving technological advancement is possible for Africa, mainly because there was a time when the continent led the rest of the world in technology and civilisation.

The book aims to provide technology stakeholders and citizens with relevant information needed to position Africa in the global tech race to unleash a wave of economic development that would lift Africans out of poverty and re-establish the continent as the worldwide hub of technology and civilisation.

Africa needs to embrace the Fourth Industrial Revolution and the technologies fuelling it. Currently, the continent is not at par with the rest of the world – Europe, the United States, China, Japan, UAE, and others. However, if we are willing to invest the time and resources needed, Africa has what it takes to leapfrog and catch up with the developed world as quickly as possible.

Africa has a large population, mostly of brilliant young people, to whom the emerging technologies of the Fourth Industrial Revolution are well suited. Given this demographic advantage, it goes to reason that the African continent can play a significant role in the emerging industrial era if the leaders, heads of governments, and other relevant stakeholders are firmly committed to facilitating and encouraging this lucrative form of development - in technology and innovation.

Even at the continental level, I am optimistic that the African Union (AU) would make concerted efforts to ensure that the entire continent is united in pursuing this aspiration of a technologically developed Africa. However, what prevails is a situation whereby most African countries have an intra-connection within their respective domains but not with other African countries. It would be best if Africa is interconnected and united in promoting the adoption, diffusion, and active development of digital technology in Africa.

The book has 21 chapters. Its first chapter puts things in perspective by taking the reader back to the basics with the topic, "African Technology in Antiquity." The idea is to immerse the reader in the storyline of once upon a time when Africa was ahead of the rest of the world in technological development. This is no surprise, given the well-documented historical, archaeological, and scientific expositions that show that Africa was indeed the cradle of civilisation and the starting point of humanity.

Chapter 2 discusses issues in the literature on development and its opposing side, underdevelopment. It opines that Africa is today the poorest continent in terms of modern development, even though it is one of the richest in terms of natural resources and human resources or demographics. It places most of the blame for Africa's current underdevelopment squarely on colonialism, slavery, neo-colonialism, leadership myopia and greed.

As this is primarily a book on Information Technology (IT) and Africa, chapter 3 explains most of the concepts and current issues in the domain to assist readers who do not understand Information Technology from a conceptual and technological point of view to do so and benefit from subsequent chapters fully.

Chapter 4 looks at computer hardware development in Africa and how Africa can concentrate on developing its hardware to meet its local demands and compete with more developed countries. The chapter looks at how African hardware startups ramp up production of everything from agricultural innovations to automobiles.

Chapters 5 through 8 detail the areas of IT that have the most far-reaching effects on the different sectors of African economies. Hence, it presents the concept of exponential technologies, the critical technologies in the domain that are exponential in their growth, and their potential impacts on growing economies. As Artificial Intelligence is today perceived as the new electricity in terms of its potential impacts, it is given a more detailed treatment for the readers in Chapter 6.

Chapter 7 looks at the megatrends we can expect in the future through prognostications. Chapters 9 to 12 cover the most potential areas where exponential technologies can be deployed to engender the best development for African countries. They also look at the socio-cultural implications of the technologies in Africa and its people and institutions. As the book sees these technologies to be necessary for African development, it also looks at how we can enable Africa to become an IT powerhouse by not only ensuring that IT and the exponential technologies are heavily diffused in the different sectors of the continent's economy and made excellent use of, but also ensure that Africa participates effectively in the innovations that will occur in many areas of IT going forward.

Chapter 13 aims to assist individuals and organisations with the knowledge and know-how on having an organised and structured methodology for systems procurements for use in organisations and how new systems can be designed and developed. Chapter 14 discusses how technologies often wear the cultures of the countries that produce them and, thus, suggests means to indigenise the IT systems we use in Africa.

Chapters 15 to 17 investigate IT in the corporate environment. Organisations need to ensure that they have a viable IT strategy plan that is regularly repeated. Organisations equally need to ensure that boards are made more effective using IT. How IT departments can be organised to produce the most effective and efficient results are also investigated. Countries need to ensure that they have a sound IT strategy. Chapters 18 and 19 look at national and continental level IT strategies and development issues in how nations within Africa need to collaborate by having a continental set of IT policies, strategies and coordinated approaches to developing IT to assure rapid development that mutually benefits all of them.

Due to the importance of youths in the usage and development of IT worldwide, Chapter 20 surveys youth participation in harnessing IT in Africa. Finally, Chapter 21 examines what universities in Africa need to do through teaching, research, curriculum design, and practical training to provide the continent with the required human capital to drive the development of this technology.

The chapters are thus written in a manner that gives the reader the choice of going through them chronologically, chapter by chapter, or independently, by choosing any of the chapters he wants to read at any time.

In order to make each chapter self-sufficient in the amount of information they provide, some of the points mentioned earlier are considerably revised. This is to equip readers with all the detailed information they need from the respective chapters, even if they had not read the previous chapters.

Finally, readers should note that the book takes a panoramic view of all the pertinent issues involved, e.g., Africa's development, IT, how to use IT in organisations, and ensuring that Africa develops more rapidly using IT. In other words, it is a breadth-first look at each of the issues, even though most of them can be treated in a depth-first manner, enough to warrant writing a whole book on each of them. The breadth-first handling of each of the issues covered in the chapters is to take the reader on a journey about technology and Africa and how Africa can benefit from it.

Evans Woherem, PhD.
Abuja, Nigeria
November 2022

Chapter 1

African Technology in Antiquity

Africa is considered the "cradle of humankind," where the first hominids and early humans evolved. This notion may be traced back to Charles Darwin. He conjectured in his 1871 book, *The Descent of Man,* that it was "likely" that Africa was the birthplace of humans since our two nearest living counterparts resided there: chimpanzees and gorillas. About five to ten million years ago, the ancient human-like fossils unearthed in South Africa led paleoanthropologists such as Robert Broom, John T. Robinson, Raymond Dart, Phillip Tobias, Ron Clarke, and Lee Berger to believe that humans originated in Africa and migrated to other parts of the globe. The continent was once referred to as Alkebulan, the "mother of mankind" or "garden of Eden", in Cheikh Anta Diop's *Kemetic History of Afrika*. However, today it is known as Africa, a continent of 54 countries with a rich mix of people, cultures, languages, economies, and histories.

It is estimated that the African population was around 1.4 billion as of 2021 (Worldometer, 2021), which is approximately 1.7 billion when added to the population of Africans in the diaspora. With a total land area of approximately 11 million square miles (about 17.6 million square kilometres), it is the second-largest continent in the world. It hosts a large conglomeration of ethnic nationalities totalling approximately 3,000. Despite this high number of diverse ethnic nationalities, Cheikh Anta Diop, a Senegalese polymath, who gained his PhD in Sorbonne, France, researched the origins and the cultural past of the peoples of Africa and concluded that they all shared a common culture and that their languages had a common origin. He equally argued during his lifetime that "there was a shared cultural continuity across African peoples that was more important than the varied development of different ethnic groups shown by differences among languages and cultures over time" (Diop, 1989).

Sankofa Bird: A Metaphor for Rediscovering the Past to Move Forward

The concept of "Sankofa" emanates from the Akan people of Ghana. It means there is nothing wrong with returning to the past to retrieve what is attainable from it that is useful to one today and in the future. In other words, we should not consider it unacceptable to go back to our beginnings or roots to be empowered or energised to move forward (Urban Dictionary, 2020). We should, therefore, be willing to trace our past to pick up the best of what we achieved – or the best that our history can teach us to reach our full potential further. This way, we will regain or resuscitate whatever we have lost, forgotten, or abandoned.

Studying African technology in antiquity shows that Africa was the cradle of civilisation. At one time in human history, the continent was more technologically advanced than the rest of the world. Unfortunately, Africa lost this pre-eminent position due to many factors, including slavery and colonisation.

Since then, African countries have lagged behind the rest of the world and have perpetually suffered underdevelopment, forcing their citizens to live substandard lives and not fully maximise their potential. In most African countries, there is a glaring lack of infrastructure, total neglect of public utilities, blatant corruption, perpetual mismanagement of public resources, and a near absence of any incentive for technological development.

Today, the technological and developmental gaps between Africa and the developed parts of the world have widened significantly. In contrast, Africa abandoned its technological trajectory and regressed in that regard; western countries gradually gained a foothold in technology and innovation, spurring their economic development. (This is because Africa missed out on the previous industrial revolutions. Today, the world refers to them as the first, second, and third industrial revolutions. In chapter 8, details of how the first industrial revolution was started in Europe and the role Europe played are provided. The chapter also points out the reason for the second and third industrial revolutions and how Africa didn't participate in them. Since the Industrial Revolution, the first, second, and third have all been pioneered in the West, if correctly motivated, Africa can be part of the fourth Industrial Revolution which is ushering in this Age of Abundance.)

Since Africa was pre-eminent in civilisation and technological development in ancient times, countries in Africa need to retrace that era to recapture the spirit of creativity, innovation, and technological advancement that once characterised and propelled it to lead in science and technology, and restart a technology-driven economic development - it is, therefore, a "Sankofa" moment for Africa.

Sankofa, perpetuated in the present age when the fourth industrial revolution is just emerging, will drive Africa from a developing continent to a developed one. The fourth industrial revolution, powered by exponential technologies, offers the best opportunity for Africa to leapfrog its economic development to compete with the rest of the world. Thus, each African country needs to pursue ICT with the seriousness it deserves, as it will catalyse the development of other technologies and all the different sectors of their economies.

African Technology in Antiquity: Contributions of Africa Towards Technology

As much as Africa lags behind in technological advancement today, this has only sometimes been the reality. In the past, Africa was way ahead of the rest of the world in civilisation and technological progress and made countless contributions in various fields of technology and human development. Archaeological findings have shown that the world's oldest records of human technological achievements are buried in Africa in the Great Rift Valley (Sah & Shah, 2008). The oldest stone tools were found in East Africa.

Also, there are many other prehistoric examples of the production and usage of tools in different regions of Africa. Archaeologists believe that Homo Ergaster produced stone tools, like axes, in Africa as way back as 1.5 million BC. By about 800,000 BC to 600,000 BC, the innovation in stone tools technology spread to other continents, like the Middle East, Asia and Europe. From about 90,000 BC to 60,000 BC, modern human beings, otherwise referred to as Homo Sapiens

Sapiens, developed bone tool technology in the Southern and Eastern parts of Africa (Shillington, 2005).

In addition, a 2017 article by Veerle Rots showed that stone knapping innovations were the case during the middle of the Stone Age in places like Southern Africa. This helped develop weapons that could be thrown as projectiles in hunting and fighting (Rots et al., 2017).

The existence and survival of humans depend on science and technology, which is an endogenous process. According to several sources, endogenous science was widespread, with women in charge of the African pottery and iron smelting when West Africans began smelting iron in approximately 400 BC (Carr, 2017). Between 1,500 and 2,000 years ago, advances in metallurgy and tool-making in Uganda, Rwanda, and Tanzania were so sophisticated that they surpassed what was obtainable in Europe. Ancient furnaces in Tanzania could reach 1,800°C — up to 300°C warmer than the Romans' (Revocatus, 2019).

Unfortunately, most discussions on the development of science and technology are today centred on the Greeks, Romans, and the rest of Western Europe, without mentioning Africa. Nevertheless, history has shown that there were remarkable technological achievements in Egypt and other African countries south of the Sahara in the past. Contributions Africans made in the past in various fields toward the development of technology include the following:

Mathematics

About 35,000 years ago, Egyptians wrote textbooks on math that explained many of the concepts taught in secondary schools today. The books dwelt on the division and multiplication of fractions. They also included geometric formulas for calculating angles, distances, areas, and volumes of shapes. They also assisted in solving problems to do with algebra and making quantitatively sound predictions on, for example, the flooding of the Nile. As early as 8,000 years ago, the Yoruba people of present-day Nigeria developed a number system that scholars have lauded as requiring more abstract reasoning to identify various numbers, and according to Blatch (2013), the system had twenty units as against the standard ten units-based framework we are used to today.

Astronomy

Ancient African cultures made several discoveries in astronomy which have remained foundational to the knowledge of the field up to the present day. The Egyptians, for instance, plotted the movements of the sun in the Milky Way and other bodies in the heavenly constellations roughly 5,000 years ago, as well as the moon's motion around the earth. They split the year into 12 parts and developed a year-long calendar containing 365 ¼ days. They also made running water clocks of the outflow variety, stone jars with sloping sides that allowed water to flow continuously from a hole drilled towards the bottom. In the Temple of Amen-Re at Karnak, a water clock was first documented in 1417 BC – 1379 BC, during the reign of Amenhotep III (Cotterell & Kamminga, 1990). Meanwhile, the African Stonehenge, a structure constructed around 300BC and found in present-day Kenya, was an extraordinarily accurate calendar.

Furthermore, the Dogon people of Mali, West Africa, were well aware of the Sirius star system and its orbit, the rings around Saturn, the moons that circumnavigate Jupiter, and even the spiral nature of our galaxy, the Milky Way (Blatch, 2013). Also, hundreds of years ago, the Dogon of West Africa knew that our galaxy, the Milky Way, had two known stars, the sun and another one that is not too easy to see with the naked eye, despite its vast density, which we now know as Sirius B.

Architecture/Engineering

Africa's most renowned architectural wonders are those found in ancient Egypt, including the monumental obelisks, the ancient pyramids, and the Great Sphinx, built between 2,000 BC and 100 AD. Besides the architectural wonders of Egypt, there were also grand buildings in other parts of Africa. In the 13th/14th century, Timbuktu, part of the great Mali empire, had spectacular university buildings, royal palaces, great mosques, and other architectural wonders. Up till today, the construction of the mosques is still perceived as primarily original, showing the use of ancient techniques.

As early as 1000 BC, the Asantes of modern-day Ghana and Ethiopians built tombs and other monuments with carved stones. It was not uncommon to construct an entire temple with stones cut from a single block of rock. Tigray in Ethiopia bears many examples of these. The ancient Benin Kingdom is perceived as having the biggest earthwork in the world; the Benin Wall was built before modern technologies were used in construction (Guinness Book of Records, 1974 Edition). It is also affirmed to be longer than the Great Wall of China by at least four times.

Navigation

It is believed by many that European navigators were the first to go by sea to the American continents. This has been proven not to be precisely correct. Evidence incontrovertibly indicates that back in 445 BC, the Greek historian Herodotus described King Ramses III commanding a crew of Africans at sea with incredible nautical and navigational skills. Also, Harvard University's Leo Weiner offers one of the most compelling pieces of evidence to support the claim that Africans sailed boats to the continents of America and Asia hundreds of years before Europeans arrived. According to Weiner's book, *Africa and the Discovery of America*, Columbus reported in his journal that "black-skinned people had arrived from the southeast on boats, bartering in gold-tipped spears." Both accounts suggest that Africans sailed to America before European navigators. Descriptions and genetic evidence from plants and art forms from South American cultures point to a movement of some individuals from West Africa to the east coast of South America. Africans could navigate to Asia and the Americas with the help of different sizes of boats that they constructed, and also as a result of the knowledge they had about rivers, oceans, navigation and how to cook and fend for themselves over a long period. The people of Mali and Songhai Empires built sizeable ships with depth of about 100 feet and a width of about 13 feet. These boats were known to carry loads of up to 80 tons (Utsua, 2015).

Textile

From early times, Africans displayed great skill in textile innovation, particularly developing indigo dyeing and dye pits using locally-produced cement.

By 5000 BC, Egyptians were cultivating flax to weave linen. By 2,000 BC, archaeologists discovered diagrams of early looms in the tombs of Egyptians, along with remnants of linen materials. The initial looms they constructed were used on the ground, without heddles. Later, looms with single heddles, which two people operated on, were found in the tombs. Subsequently, threads with single heddles, which could be vertically mounted and secured against a tree, were produced (African Textile History, 2016). In the Senegambia region of West Africa, cotton was spun into yarn and woven into cloth from as early as 1000 AD, long before Europeans arrived on the continent.

Textile production flourished in other African regions, such as the specimens of ancient multi-coloured tunic fabrics found in Niger, identified as going as far back as the 8th century, and woven fibres dating to the 9th century, found in present-day Nigeria. Also, an 11th-century woven cotton cloth was unearthed in Mali (African Textile History, 2016).

In the ancient city of Meroe, which was located on the east bank of the Nile, the Nubians who lived there had a reputation for producing quality woven textiles. The Cameroonians of Central Africa were famous for making fabrics from the bark of trees, while other African regions used animal hides, furs, and even feathers for the same purpose.

Metallurgy and Tools

Evidence from Djenné-Djenno, considered one of the oldest urbanised centres and the best-known archaeology site in Sub-Saharan Africa, shows that iron production took place in the region as early as 250 BC (Becker, 2009). The Bantu of Central, Southeast, and Southern Africa spread iron production technology to those regions between 500 BC and 400 AD.

While indications suggest that the people of present-day Nigeria and Cameroon may have practised iron smelting from around 2,000 BC, there is much evidence from traditional oral accounts that indicate that the Nok culture of present-day Nigeria carried out iron works even as far back as early 1,000 BC (Wikipedia contributors, 2022).

From 400 BC to 200 BC, shaft furnaces with many bellows and huge inner diameters of about 30 to 45 inches existed in different African regions. Following this initial technology, Africans developed a more sophisticated range of furnaces. For example, the tiny slag-pit furnace was developed in Senegal from about 400 AD – 600

AD. Africans also developed "21-foot-tall natural draft furnaces in 20th century West Africa" (Hirst, 2008). Some of the furnaces had a permanent shaft; others had a movable, portable shaft, while others did not have a shaft.

Across Sub-Saharan Africa, evidence abounds of regions and territories using natural draft furnaces for iron production. For example, in Togo, iron production reached some 80,000 cubic metres (Shedreck, 2015). In other parts of the continent, slag production reached 169,000 cubic metres and 300,000 cubic metres in Burkina Faso's Korsimoro and Fiko districts, respectively (Killick, 2015).

Besides their expertise in iron smelting, ancient Africans also used bronze and brass to make internationally-recognised objects. For example, in the 13th century, a lifelike statue made of brass was found in Ile-Ife, Nigeria's ancestral home of the Yoruba ethnic group.

Medicine

Long before Europeans appeared on the continent, Africans were using many of the treatments we use in medicine today to cure illnesses. For example, across present-day Nigeria, Egypt, and South Africa, Africans use kaolin to treat diarrhoea and plant extract to treat pain, kill bacteria, cure malaria and manage cancer.

During the same time, Africans discovered and used ouabain to treat supraventricular arrhythmias and congestive heart failure. It was also effectively used for treating chronic atrial fibrillation, where the substance was employed primarily to control ventricular rate. Also, they used the substance as poison for arrows. These treatment regimens were as effective as what modern Western medicine offers.

In North Africa, particularly in present-day Tunisia, ancient Africans extensively used a plant known as wormwood to treat various health challenges such as diabetes, hypertension, neuralgias, and bronchitis. In addition, they made herbal teas with antibacterial properties from wormwood's bark, which they employed in treating various ailments.

In Botswana, Namibia, and South Africa, a plant called the Devil's dew has been used to cure blood diseases, diabetes, gastrointestinal disorders, allergies, ulcers, and skin injuries for centuries. It is also efficacious for treating ailments like headaches and joint and back pains.

In Madagascar, the plant known as periwinkle was of great medicinal value. Africans traditionally used it to treat venereal diseases, skin disorders, and rheumatism. They also leverage it for its efficacy in lowering blood sugar and glucose levels to treat diabetes and other blood-sugar-related ailments. In Uganda and some other parts of Eastern Africa, caesarean surgeries were successfully practised with anaesthetics on pregnant women without complications and using banana leaves to relieve pain.

All the above antiquities show that technological prowess is not new to Africans. Africa was indeed a pioneer in various areas of technology. However, unfortunately, this mastery and leadership were lost mainly due to colonialism and Western hegemony, which is currently perpetuating a largely detrimental neo-colonialist agenda in Africa and, of course, through a set of ineffectual leaders who took over from the colonial masters. The next chapter takes a more detailed look at issues to do with development and underdevelopment so that readers can see when the rain of underdevelopment started beating Africa and what could be done to assuage the current state of underdevelopment.

References

African Textile History. (2016, August 4). Retrieved August 26, 2022, from https://study.com/academy/lesson/african-textile-history.html.

Angelis Revocatus. (2019, March 24). *What are some of the greatest contributions of Sub-Saharan Africa to human development?* Quora website. Retrieved August 26, 2022, from https://www.quora.com/profile/Angelis-Revocatus

Becker, J. (2009, February 17). *Djenné, city of clay - World Heritage*. UNESCO Multimedia Archives. Retrieved August 26, 2022, from https://www.unesco.org/archives/multimedia/document-89

Blatch, S. (2013, February 1). *Great achievements in science and technology in ancient Africa*. American Society for Biochemistry and Molecular Biology. Retrieved August 26, 2022, from https://www.asbmb.org/asbmb-today/science/020113/great-achievements-in-stem-in-ancient-africa

Carr, K. (2017, May 19). *African Science and African Inventions* - History of Science Quatr. Us Study Guides. Retrieved August 26, 2022, from https://quatr.us/history/african-science-history.htm

Cotterell B. & Kamminga J. (1990). *Mechanics of pre-industrial technology: An introduction to the mechanics of ancient and traditional material culture.* Cambridge University Press.

Diop, C. A. (1989). *The cultural unity of Black Africa: the domains of patriarchy and of matriarchy in classical antiquity.* London: Karnak House.

Hirst, K. K. (2008, September 9). *African Iron Age - 1,000 Years of African Kingdoms.* ThoughtCo. Retrieved August 26, 2022, from https://www.thoughtco.com/african-iron-age-169432

Killick, D. (2015). Invention and Innovation in African Iron Smelting Technologies. *Cambridge Archaeological Journal, (25(1), 310 - 316)* Cambridge University Press.

Rots, V., Lentfer, C., Schmid, V. C., Porraz, G. & Conard, N. J. (2017). Pressure flaking to serrate bifacial points for the hunt during the MIS5 at Sibudu Cave (South Africa). *PLoS ONE,* 12(4): e0175151. https://doi.org/10.1371/journal.pone.0175151

Sah, A., & Shah, M. (2008). The great rift valley of East Africa. Cape Town: Struik.

Shedreck, C. (2015). Metals in Past Societies: A Global Perspective on Indigenous African Metallurgy. *MRS Bulletin* 40, 88. https://doi.org/10.1557/mrs.2015.251

Shillington, K., (2005). *History of Africa*, Rev. 2nd edition. New York: Palgrave Macmillan. Retrieved August 26, 2022, fromwww.africaworld.net/afrel

Urban Dictionary, (2020, September 30). Urban Dictionary: Sankofa. Urban Dictionary/ https://www.urbandictionary.com/define.php?term=Sankofa

Utsua, P. T. (2015). The Evolution of African Indigenous Science and Technology, in Historical Research Letter, 16, 14-21. Retrieved August 26, 2022, from https://iks.ukzn.ac.za/node/1067

Wikipedia contributors. (2022, June 26). *Iron metallurgy in Africa.* Wikipedia. Retrieved August 26, 2022, from https://en.wikipedia.org/w/index.php?title=Iron_metallurgy_in_Africa&oldid=1095175281

Wikipedia contributors. (2022, May 3). *Economy of Africa.* Wikipedia. Retrieved 10:11, May 6, 2022, from https://en.wikipedia.org/w/index.php?title=Economy_of_Africa&oldid=1086025071

Worldometer. (2020). *Population of Africa (2019).* Worldometer - Real Time World Statistics. Retrieved September 2, 2022, from https://www.worldometers.info/world-population/africa-population/

Signé, L. (2018). *Capturing Africa's high return.* Brookings Institution. Retrieved December 6, 2021, from https://www.brookings.edu/research/africas-consumer-market-potential/

Chapter 2

Under-Development of Africa Since Colonialism

The concept of national development has proven challenging to define. Since the 20th century, various scholars have tried to distil the theories around it into a workable definition. However, academics have criticised such efforts, rivalling those of the definition's proponents.

Nevertheless, most economists and sociologists now agree that development should improve the well-being of society. Hence, Seers (1961) defined development as "...the reduction and elimination of poverty, inequality, and unemployment within a growing economy."

For a nation to be perceived as developed, it must also enjoy significant economic growth. An economy grows if it records a sustained increase in the value of wealth it produces over a given period. One measure of wealth is the Gross National Income (GNI), the total value of products and services made by a country's citizens, regardless of where they are. Growth in individual productivity and wealth is measured by Per Capita Income, which is calculated as the GNI of a country divided by its total population.

Following this measurement, countries may be placed in one of three categories: Lower-Income, Middle-Income, and High-Income. These classifications overlap with the concept of underdeveloped and developed countries. First, most underdeveloped countries are Low-Income or Lower Middle-Income countries; High-Income countries are usually classified as developed countries. A country is High-Income and developed if its GNI Per Capita (Per Capita Income) exceeds $12,056.

Most countries in the Higher-Income category are in one of these regions: North America, Europe, or the developed parts of Asia and the Pacific. The UN notes that, as of 2020, about 35 countries might be considered developed (UNCTAD, 2022). The High-Income, developed countries include the United States and Canada in the western hemisphere; France, Germany, Spain, the Republic of Ireland, Italy and Switzerland in Europe; and Australia and New Zealand in Oceania.

However, it is recognised that generalised economic growth does not guarantee individual prosperity. Income disparities, measured in the Gini coefficient, show that wealth created by labour and technology is often unevenly distributed. Therefore, if development efforts are to succeed, there has to be a conscious effort aimed at "the reduction and elimination of poverty" and not merely at increased GDP and per capita income.

For actual development to occur in a society, there has to be an increase in the technical capacity of its people; in other words, their ability to produce more excellent quality products and services within a given period.

The United Nations Development Index (HDI) measures a country's human development level compared to other nations. The HDI considers things such as the level of education, living standards, and life expectancy of a person. The measure is in line with the UNDP's definition of development, which involves enlarging the choices a people can make in their environment, chief of these choices being "to lead a long and healthy life, to be educated and enjoy a decent standard of living" (UNDP, 2008). Interestingly, the top countries by HDI also appear on the list of High-Income countries, albeit positioned differently. Those countries are "Norway, Ireland, Switzerland, Hong Kong, Iceland, Germany, Sweden, Australia, Netherlands, and Denmark." Though highly developed countries, the United Kingdom and the United States rank 13th and 17th, respectively, on the list.

Development is said to have occurred when there is a manifest level of enhancement in the quality of life in a country. A society will achieve its growth and developmental aims if its people are literate, better nourished, and educated. Education enables them to develop and use technologies that increase wealth creation and income.

What is the Underdevelopment of a Country?

An underdeveloped country has a low level of development, as evidenced by its low per capita income, high poverty rates, low literacy rates, low life expectancy, and inefficient resource utilisation. In underdeveloped countries, most of the population has a low quality of life (Wikipedia contributors, 2022a).

Underdeveloped countries may be found in Africa, Asia, the Pacific, and the Caribbean. However, most of the UN's least developed countries are in Africa. This is because they differ somewhat from what economists call "developing countries," which are at various stages of transition away from underdevelopment and toward development. Underdeveloped countries include Niger, Sierra Leone, Uganda, Rwanda, the Democratic Republic of the Congo, Zambia, Sudan, Haiti, Myanmar, and Nepal (Wikipedia contributors, 2021).

These states are overwhelmingly Low-Income (with a per-capita income of less than $1,026) or Lower Middle-Income countries (Per Capital Income between $1,026 and $4,038). In material terms, the income disparities between underdeveloped and developed countries can be vast. For example, the average income of the United States (GDP per capita of $65,298; GDP per capita of $775) is *84 times* that of Sudan (GDP per capita of $775) (Wikipedia contributors, 2022b).

The most visible characteristic of an underdeveloped society is the prevalence of absolute poverty. Absolute poverty is the inability to adequately meet basic human needs: food, clothing, and shelter. For this mass, the primary goal of everyday life is survival. Another feature of underdevelopment is a relatively low level of education. As a result, many people in these countries are uneducated or not educated enough to begin or sustain the process of capital formation and the building and use of technologies required to create economic wealth.

Countries that are found in this category typically have low HDI. For example, the countries in the bottom ten HDI rankings have an average HDI score of 0.459. Compare that with the top ten, which have an average score of 0.949. The picture is clear: the gulf between developed and underdeveloped countries can be pretty stark regarding human capacity development.

In practical terms, underdeveloped countries cannot fully exploit the resources at their disposal to their advantage. These resources include material and human resources. This is directly linked to the suboptimal levels of education and their insufficient or non-existent transfer of capacity and technology to underdeveloped nations from countries that possess it.

Despite widespread agreement on some of the features of an underdeveloped country, the term arose from a historical dispute about the nature and causes of the deprivation it describes. For much of the study of development in the West, countries in this category had been represented as "developing," meaning that these nations were in a transitory stage on the development path. It was thought that such a path would lead to a destination (development) that mirrored a reality much like Western nations' economies, essentially possessing Western technologies.

Controversy over the division of the world into two stratifications (developed and underdeveloped/developing) has caused some recent revisions of the terms involved. For instance, in 2015, the World Bank announced that it would no longer use the developing/developed distinction. Instead, it would simply present aggregate data from the various regions of the world (Khokhar & Serajuddin, 2015). However, this step does nothing to resolve the negative realities characterising underdeveloped societies.

In his 1972 book, *How Europe Underdeveloped Africa*, Walter Rodney, a Guyanese historian and academic, described Africa as underdeveloped rather than developing. He believed that European colonialists had caused a deprivation of economic conditions in Africa by exploiting the continent's resources for their gain (development). According to Rodney (1972), this was done at the expense of native Africans. Thus, his theory of underdevelopment did not just describe a fixed state of affairs; it also captured a continuing process by which Africa was subjugated and prevented from developing by Western powers.

How Colonialism Underdeveloped Africa

Walter Rodney's thesis was one of several works advocating the dependency theory by Marxist intellectuals. The dependency theory holds that resources flow from poorer states on the periphery of the global economy to wealthier ones at the centre, thus intensifying the poverty of the poorer states and the wealth of the richer ones. In other words, (predominantly Western) states need to exploit poorer countries to remain prosperous. This need fuelled the West's colonial activities in Africa, where they exploited the resources of the countries they conquered and exploited the local populations as cheap labour. These wealthy countries also made their colonies the final markets for their products. Thus, colonialism led to Africa's underdevelopment.

Dependency theorists like Walter Rodney argued that to grasp the nature of the relationship between Africa and its Western colonisers, three themes had to be understood:

1. The development of Africa in pre-colonial times;
2. The developmental condition of Europe before colonial expansion; and
3. How the subsequent colonial contact between Africa and Europe contributed to each other's present condition.

This school of thought saw events such as the Scramble for Africa (the partitioning of Africa by European powers) as key to the current underdevelopment of the African continent. Before this, African states developed at a pace suited to their local, physical and cultural environment. Their technologies, though not perfect, enabled them to exploit resources to enhance their communities. Trade between states and regions was expanding.

The arrival of colonialism disrupted and reversed many of these gains. The goal of colonialism was the acquisition of Africa's resources by European masters at the lowest possible price for the coloniser's economic benefit. Europe was still developing, so they reaped African raw materials for this purpose. They enlisted Africans to do the hard work of mining the resources for low wages. The infrastructure and technologies they built were mainly for conveying these resources to the ports, from whence they shipped them to the West.

With local African economies now focused primarily on providing resources for the West, there was little chance of building infrastructure and technology for other purposes—certainly not for growing and developing the local economy.

The African colonies became dependent on their First World colonisers on multiple fronts, including economics, technology, politics, education, finance, media, and the development of human resources. First World nations achieved this by setting the terms of international trade. While poorer countries were not detached from global business, their participation in it was determined by colonial First World states, which restricted the export and import of particular items to and from their colonies.

According to Rodney (1972), Africa became subject to the will of wealthier nations because it lost the power to determine its fate. He explained that power was "the ultimate determinant" of the relationships between people within and between groups. The latter applies to the engagement between Africa and its would-be colonial overlords. Because Africa lost its power to them, it could not assert itself against the exploitative tendencies of colonisers. By relinquishing this power, Africa became susceptible to the underdevelopment engineered through colonialism.

Walter Rodney noted that underdevelopment was not a lack of resources but the uneven distribution of resources, which benefitted exploitative classes and states at the expense of poorer nations. This uneven distribution was enabled by policies drawn up by wealthy states, although they were not always deliberately intended to disadvantage countries at the periphery of global economics.

Another contributor to the school of dependency theory, Frank (1966), rejected the idea that countries were underdeveloped because they were isolated from the global economy (a claim made by Western scholars of the neoclassical economic tradition). Instead, he said, they suffered because developed countries exploited poorer countries, a situation mirrored by a metropolis-satellite relations model.

He rejected the suggestion that underdeveloped countries could learn from the experience of the now-developed countries in Europe and North America, as countries in these regions had industrialised partly by exploiting other countries and making them their "satellites" (dependents). For example, Frank (1966) pointed out that Brazil had a relatively high rate of industrialisation soon after World War II but predicted that it would remain dependent on the West because it relied on Western countries as its chief export markets. As a result, they could set the terms of trade and keep countries like Brazil at a disadvantage.

Chinweizu Ibekwe (1987), also a scholar of the dependency theory, zeroed in on Africa's state as a conquered region as a critical factor that drove its underdevelopment. He blamed Africa's inability to defend itself from conquering powers for its present state. This historical failing, he suggested, had followed from centuries of unequal engagement with the West and culminated in Africa's eventual capitulation to the latter. Chinweizu (1987) also focused on the part played by Africans who had become willing partners of the West in the subjugation of their continent. He contended that without attention to the ideology that spurred these people's actions, observers would fail to understand why an African would want to collaborate with his oppressors. Their actions enabled and sustained Africa's depressing status quo. Through his critique, Chinweizu recognised the ideological dimension that informed the persisting underdevelopment of Africa. His book, *The West and the Rest of Us*, expounded upon these ideas.

In what scholars call neo-colonialism, the operating structures built by colonisers remained long after colonialism ended. This refers to the continuing dependence of African states on western nations after colonialism had supposedly ended. Although former colonies had declared independence from Western overlords, they still functioned along the lines defined by colonisers. In other words, their production, development of labour and technology, trade, education, and politics are all subject to the needs or influence of Western powers.

A major proponent of the concept of neo-colonialism was Kwame Nkrumah, a former President of Ghana. He explained that Western nations had replaced colonialism with neo-colonialism, as they were utilising aid and investment in former colonies in ways that widened the economic gap between the former colonies and their ex-overlords.

How African Leaders Have Contributed to the Underdevelopment of Africa

Given the era when Rodney wrote his book, it is impossible to argue with his assertions of how Europeans exploited African countries. The book was written barely 19 years after many African nations gained independence from their colonial masters. Some African countries were still under colonial rule, including Zimbabwe, Mozambique, Angola, and Namibia. To that extent, Rodney's claims in the book were mainly valid.

However, it has been over 50 years since the publication of *How Europe Underdeveloped Africa*, so it does not seem wise to heap the blame for the underdevelopment of Africa today solely on the shoulders of our colonialists. If Rodney had lived up to the present, he would have developed a different perspective from the one in his famous book regarding the underdeveloped status of the African continent up to the present.

Many African countries have gained independence in the last 60 years, a duration long enough for any country to make significant economic and technological progress. However, countries in this category, primarily within Sub-Saharan Africa, have remained in a state of underdevelopment for the entire duration of their independence.

It has become apparent that the underdevelopment of Africa stems primarily from corrupt leadership on the part of heads of government and others who have had the opportunity to occupy leadership positions in various African countries. These individuals have failed to properly channel the resources – natural, financial and human – of Africa to foster development in their respective countries. Instead, they embezzle public funds meant to fund infrastructural and other developmental projects and funnel them to their bank accounts in Western countries for safekeeping.

According to the African Union Advisory Board Against Corruption (2020), an estimated sum of $148 billion meant for public development in Africa is embezzled annually by public office holders, civil servants, executives in multi-national corporations, and business executives in collaboration with executives of financial institutions in Europe where the monies are stashed as the personal funds of those corrupt leaders.

In Nigeria, former Chairman of the Economic and Financial Crimes Commission (EFCC), Nuhu Ribadu, alleged that since the country became a republic in 1963, about $400 billion had been looted from the public treasury by various Nigerian leaders (Ademola, 2011). In his book, *Africa Unchained: The Blueprint for Africa's Future,* George B. N. Ayittey gave a breakdown of funds stolen by some African heads. The sum of the amount came to a whopping $37.683 billion (Ayittey, 2005).

From the foregoing, it is clear that corrupt leadership has contributed significantly to the underdevelopment of Africa. As a result, there has been a glaring lack of sustainable development programmes that meet the people's aspirations, particularly concerning lifting most citizens out of poverty. If the funds that various leaders reportedly stole had been invested in building factories, schools, hospitals, infrastructure, and funding technological development and innovation, the economic state of the continent would have been significantly better off than it is currently.

Besides the glaring corruption that is the stock in trade of most leaders of African countries, there is another reason for the continent's underdevelopment: leadership myopia. In most African countries, the leadership lacks the capacity for strategic visioning in terms of long-term policy and decision-making, as well as implementation for the holistic growth of the respective countries. Most African leaders are so myopic that rather than promote merit, ethnic, regional and religious sentiments are influential factors that inform their decisions at the highest levels of government – including appointments to sensitive positions, award of contracts, and recruitment into various government agencies and departments.

Rather than focus on the urgent development of their respective countries, most leaders of African countries are obsessed with whose regions or localities are entitled to ascend to leadership next or which regions are more deserving (in their particularistic mindsets) to have some new projects or government positions taken to their villages, communities or regions. As these leaders carry on this way, they are oblivious to the widening gaps between their countries and other countries regarding economic and technological development.

How Technology Acquisition Accentuates Development

As it has been pointed out previously, a necessary condition for the development of societies is an increase in the quality and quantity of products and services they produce. Indices such as GNP and income per capita are often used to measure this productivity (which translates to wealth on the national and individual levels, respectively).

Underdeveloped countries score relatively low on these indices, hence their classification as low-income states.

Economists speak about the "catch-up effect," which is the idea that developing and developed economies will eventually converge in income per capita because poorer economies tend to grow faster than developed countries. However, this tends to happen if the more impoverished countries fulfil certain conditions, including the ability to attract capital, absorb and adapt technologies, and engage in global markets.

Technology, in particular, plays a crucial role in the movement of countries from the underdeveloped/developing stage to the developed stage. The greater the technological capability of a country, the higher the likelihood of the country to enjoy more significant economic and developmental strides. If sustained, this growth could lead to the creation of more jobs and, eventually, to prosperity. Technological capability is using different technologies to produce other products in increasing quality and quantities. Given the positive relationship between technical ability and economic growth, developing countries will need to progress up the technology scale if they are to increase their productivity.

Underdeveloped countries have to achieve these things in particular:

- Attain the minimum quality that lets them enter into a globally competitive market for several products;
- Spread productive capabilities across each country's working population; and
- Progressively improve quality across multiple product categories.

Technological capability is acquired not only by formal education (acquisition of scientific knowledge) but also by the propagation of tacit knowledge of production processes. Tacit knowledge is gained through learning-by-doing, i.e., hands-on experience. If developing countries lack this applied knowledge of technology, they will struggle to grow the quality and quantity of products produced per person. This is very much the case for developed countries with a large population of formally educated individuals.

Vital technological skills can only be built using the required technology over time. This, in turn, means that the technologies will have to be acquired ("absorbed") from the countries that possess them. Thus, the two vital ingredients for growth via technological enhancement are:

- Human capacity or know-how; and
- Machines and other relevant technologies.

The combination of these two factors can lead to economic development in at least four ways:

1. **Direct and Peripheral Job Creation:** The adoption of improved technologies and the strengthening of skills can lead to the expansion of existing industries and the creation of new ones. This invariably means creating new jobs and the evolution of various local adjacent/peripheral businesses. The broader economy benefits from this.
2. **Growth in GDP/GNP:** Other factors being constant, a broad-based increase in the technological capability of a country will drive growth in its total production, as measured by its GDP. A well-known example is that a 10% increase in broadband penetration in a country can contribute to a 1.4% increase in its GDP. In other words, the total product—or wealth—created by a government can increase significantly when there is an increase in the deployment of functional technologies.
3. **Workforce Transformation:** The continued use of acquired technologies by the workforce in developing countries over time should lead to a deepening of capacities that are transferable to even more people. As more people are trained in using these technologies, the workforce is transformed and becomes more competent.
4. **Innovation:** As developing countries work with technologies they have acquired, they will likely find innovative ways of applying them to their local situations. Domestication can yield technological offshoots that are more suitable for the developing communities' realities.

Japan provides an example of how technology acquisition could lead to economic development. Its economy was devastated by the Second World War. In the 1950s, it established an environment conducive to economic growth and imported machinery and technology from the West, principally from the United States. It also strove to develop indigenous technologies. These developments enabled it to manufacture products that could compete in global markets and earn from exports. By the 1960s and 1970s, Japan recorded very high growth rates and improvements across other indices. Since the late 1970s, it has ranked among the world's top five economies by size.

China, another example, has followed a U-shaped trajectory. In ancient times, it was ahead of the West regarding its technological capacity and development. It invented paper about 2,000 years ago and mastered printing hundreds of years before the Europeans did. In China, other critical technologies of the previous century were also invented. Namely, technologies like gunpowder, compasses, fired bricks, and movable parts. As late as 1700, China's GDP exceeded all of Europe *combined*.

However, after the British forcefully invaded China in the 19th century, the country fell into a period of stagnation. At the same time, Western countries powered ahead as the Industrial Revolution climaxed. China fell far behind Europe and grew slowly throughout the early and mid-20th centuries.

Closer to the present, China's economic power has been resurgent. Between 1980 and 2005, its GDP grew at an average of about 10% yearly. Technology has been a critical ingredient in the rapid growth of China. In 2020, digital technology alone contributed a third of China's national output. Over the past few decades, the country has made a concerted effort to increase its technological and industrial capacity, a move that has paid off reasonably well (UNCTAD 2022). They provided an example that African countries could imitate.

The Positive Correlation Between Technology and Development

It is generally agreed that technological capacity and development are positively correlated. This relationship also appears to be self-reinforcing. When combined with human development (i.e., when it entails both the technical know-why and know-how), technology engenders economic growth and increases human well-being.

This model comes in two parts. The first part runs from economic growth to human development. It is powered by private and public (government) investment, effort, and technology. These factors are converted into improvements in education, health, economic well-being, and poverty reduction by technology. The second part of this model runs from human development to economic growth, again mediated by technology.

We will concern ourselves with the second part of this model. It mirrors the new growth theory of development, in which technology is closely linked to education, health, and production inputs. Unlike other prevailing development theories, the new growth theory says that the desires and wants of a country's populace (and not uncontrollable external factors) drive its productivity and development. The theory emphasises the importance of a country's innovation and technology, among other factors, for its economic growth. Theorists of this school recommend that governments prioritise access to education and support private sector Research and Development (R & D), which enables its beneficiaries to meet local and global wants (and ultimately grow the economy).

However, countries must be independent of the import of technology to drive growth. Instead, they need to spend on science and technological research (R & D) to develop local capacity and sustain technology-driven growth and development.

The Vicious Cycle of Underdevelopment vs the Virtuous Cycle of Development

The concepts of the vicious cycle of underdevelopment and the virtuous cycle of development allude to a chain of factors that reinforce and intensify the status quo of economies. These factors cause underdeveloped countries to remain underdeveloped and developed economies to remain developed. Both these cycles are worth studying. They reveal the difficulty of lifting underdeveloped nations out of poverty and the overwhelming likelihood that more prosperous countries will stay rich.

The Vicious Cycle of Underdevelopment

This concept was developed by Nurkse (1961), an Estonian economist. It shows how existing disadvantages within underdeveloped countries reinforce the status quo of underdevelopment. According to the model, the accumulation of capital (investment in fixed capital, such as factories and machines) provides underdeveloped countries with an escape from poverty. However, this cannot happen because there needs to be more savings to drive the investment that creates capital. But unfortunately, savings are low because absolute income levels are low. In other words, a low-income group in a society will lead to low savings and investments. This will consequently lead to a low level of capital accumulation. This is the vicious cycle of underdevelopment.

So far, economists have described the *supply side* of the vicious cycle of underdevelopment. The equally important demand side goes thus: low income leads to low demand for goods, leading to a limited market and low investment.

Nurkse (1961) notes, “On both sides of the problem of capital formation in poor countries, a vicious cycle exists.” He says, “Because it is poor, the country does not develop; because it does not develop, it remains poor.”

Figure 1 illustrates the vicious cycle of poverty or underdevelopment.

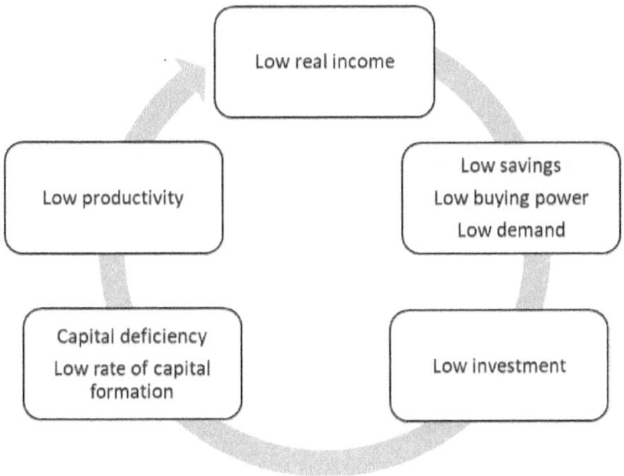

Figure 1: The vicious cycle of poverty/underdevelopment

The Virtuous Cycle of Development

The virtuous cycle of development refers to the stream of savings, investment, and productivity that reinforce positive growth and result in a developed economy. This cycle traces the positive effects that high capital formation has on an economy. High capital formation happens when incomes are high enough to allow for savings, which can be deployed as an investment to form capital.

Capital formation is a necessary (although not sufficient) condition for development.

The virtuous cycle of development is illustrated in **Figure 2.**

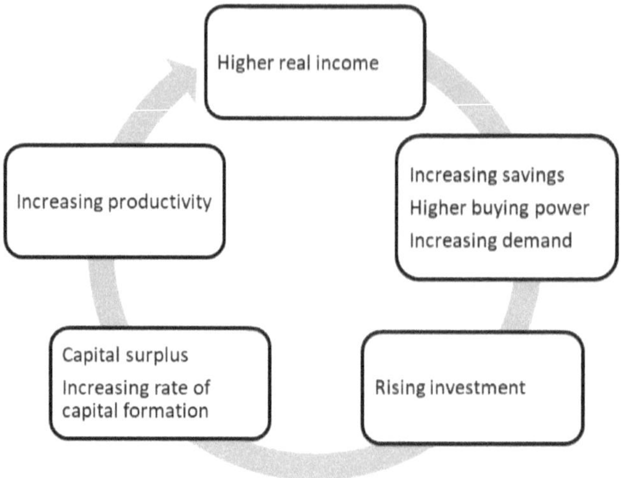

Figure 2: The Virtuous Cycle of Development

Breaking the Vicious Cycle of Underdevelopment

Experts in national development are concerned with how countries can break out of a vicious cycle of underdevelopment and begin a virtuous process of increasing growth. As the models above suggest, increasing capital accumulation is key to making this happen.

Often, the solution lies in the injection of capital and technologies from outside the economy, adapting these technologies for local use, and encouraging local innovation that enables greater productive capacity. These solutions are distilled in the four points noted below:

- Openness to international trade (modelled by exports);
- Foreign Direct Investment (FDI);
- Research and Development (R & D); and
- Various types of patents.

They are discussed below:

1. **Openness to International Trade:** Trade with more developed countries (closer to the international trade frontier) positively

impacts technological change. Developing countries can absorb technologies from their trading partners and adopt and apply them locally through trade. This reduces their need for spending on designing and fabricating the technologies locally.

2. **Foreign Direct Investment (FDI):** Multinational companies can instigate technological change by training labour, providing quality production inputs, and enabling trade connections for their host country. The multinational may start as a subsidiary, create a joint venture with local businesses, and later begin managed contracts as the host country matures. These actions help to broaden and deepen the country's technological capacity.
3. **Research and Development (R & D):** Research shows that a government's consistent spending on R & D can lead to economic growth. However, most R & D spending happens in developed countries. Developing states may take the route that Japan did early in its economic path to prosperity and sponsor R & D for vital industries instead of relying entirely on fundamental scientific advances. R & D can also be carried out to adapt foreign technologies for local use.
4. **Patents:** Patents can be an incentive for increasing technological capacity. This is because measures taken to protect intellectual property rights can enable innovative entities to reap the rewards from their investments. This is especially true for smaller-scale firms, which larger companies may otherwise outmuscle.

These four channels are known to affect economic growth positively and may lead to the improvement of development indices.

While it is essential to highlight the role played by investment in technological capacity, it is also worth remembering that technology itself evolves. This is why underdeveloped countries should focus on R & D and patents to confer themselves with some advantages over global competitors. Incentives should also be given for FDI that helps local technological progress.

It is to an examination of these newer technologies that we turn next.

References

Ademola, A. (2011). Endangering good governance for sustainable democracy:

The continuity struggle against corruption in Nigeria. *Journal of Research in Peace, Gender and Development*, 1, 307-314.

African Union Advisory Board Against Corruption. (2020). *Status of ratification of the convention on corruption*. African Union Advisory Board Against Corruption. Retrieved May 19, 2022, from https://anticorruption.au.int/en/documents/2021-06-11/status-ratification

Ayittey, G. B. (2005). *Africa Unchained: The Blueprint for Africa's Future*. New York, N.Y.: Palgrave Macmillan.

Chinweizu, I. (1987). *The West and the Rest of Us: White Predators, Black Slavers, and the African Elite*. Pero Press Lagos, Nigeria.

Frank, A. G. (1966). *The Development of Underdevelopment*. Monthly Review Press, New York.

Khokhar, T., & Serajuddin, U. (2015, November 16). *Should we continue to use the term "developing world?"* World Bank Blogs. Retrieved May 18, 2022, from https://blogs.worldbank.org/opendata/should-we-continue-use-term-developing-world

Nurkse, R. (1961). *Problems of Capital Formation in Underdeveloped Countries*. New York: Oxford University Press. p. 163.

Rodney, W. (1972). *How Europe Underdeveloped Africa*. London: Verso. p. xxii. ISBN 9781788731188.

Seers, D. (1969). *The meaning of development*. Institute of Development Studies Communication 44, Brighton: IDS. Retrieved, August 27, 2022, from https://www.ids.ac.uk/wp-content/uploads/2018/12/IDSOpenDocsStandardTermsOfUse2018.pdf

UNCTAD. (2022). *UN list of least developed countries*. Retrieved 7 May 2022, from https://unctad.org/topic/least-developed-countries/list

UNDP. (2008). Human Development Index. Human Development Reports. Retrieved, August 27, 2022, from https://hdr.undp.org/data-center/human-development-index/indicies/HDI

Wikipedia contributors. (2021, December 29). How Europe Underdeveloped Africa. Wikipedia. Retrieved 18:47, May 7, 2022, from https://en.wikipedia.org/w/index.php?title=How_Europe_Underdeveloped_Africa&oldid=1062617955

Wikipedia contributors. (2022a, March 5). Underdevelopment. Wikipedia. Retrieved 18:27, May 7, 2022, from https://en.wikipedia.org/w/index.php?title=Underdevelopment&oldid=1075458646

Wikipedia contributors. (2022b, March 17). List of regions by past GDP (PPP). Wikipedia. Retrieved 19:59, May 7, 2022, from https://en.wikipedia.org/w/index.php?title=List_of_regions_by_past_GDP_(PPP)&oldid=1077685862

Chapter 3

Understanding the Concept of Information Technology

Information technology (IT) refers to technologies that deal with human communication processes and the information they process. It is an emerging science that collects, processes, and transmits data. It explains how computers can store, analyse, and send data through various media, such as satellites, telephone lines, Teletext, and cable television.

Since writing was developed in about 3000 BC, humans have been storing, manipulating, and retrieving data. However, the use of the term "information technology" in the context of processing, storing, and retrieving data was first used by Leavitt and Whisler (1958) in their paper which appeared in Harvard Business Review in 1958.

IT comprises many different components. The first component involves processing vast quantities of data and information, which is achieved by giving computers the ability to calculate, manipulate and shuffle data and information at enormous speed. The second component uses mathematical and statistical techniques to solve problems. The third component is still being finalised; it aims to enable computers to achieve more human-like reasoning and decision-making abilities, which is achieved through the simulation of different human thinking abilities.

In present-day usage, information technology encompasses computers and computer networks as well as other technologies that distribute information, such as telephones and television (Brown, 2020; Castagna & Bigelow, 2021).

The two main areas of IT are computer systems and

telecommunication.

The computer system is useful for processing, storing, and transmitting data and information. This is further broken down into the hardware and software components. Hardware refers to a computer's physical components which it functions with. These include a motherboard, a central processing unit (CPU), ventilation fan, monitor, mouse, graphics card, keyboard, and other physical computer accessories. Software is the algorithmic and non-algorithmic instructions given to a computer system to execute tasks. It uses these sets of instructions to manipulate a computer's hardware components to carry out specified actions given to it through software codes. Such instructions are called software programmes.

Two significant kinds of software are software that act on the hardware, referred to as systems software, and software used to run tasks and processes, known as applications software. System software is sandwiched between a computer's hardware components and other software systems. It takes instructions from application software and makes the requisite hardware components carry out stipulated instructions. System software thus comprises (1) software that is used to manage the hardware components of the computer, like operating systems, device drivers and system utilities, and (2) software that is used for the development of other software, like programming languages, programmes compilers and interpreters.

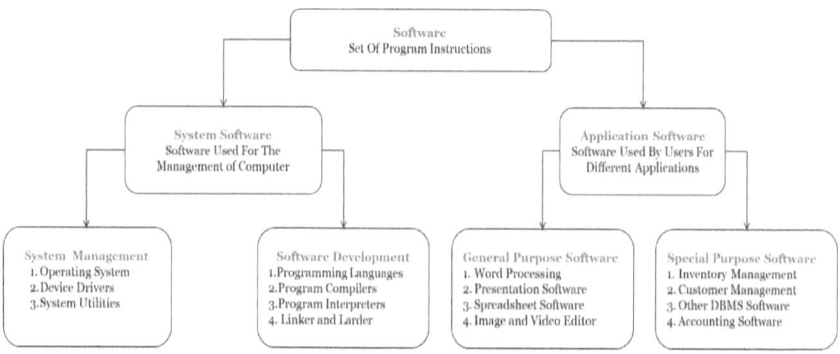

Figure 3: System and Application Software

Application software or "app" is a computer programme that is designed to perform a specific task based on a set of requirements given by an end user. Significant types of application software are spreadsheets, databases, word processors, ERPs (Enterprise Resource Planning solutions), media players, payroll systems, and others.

Many people consider telecommunications to be the most profitable business in the world. However, some believe that telecommunications exclusively deals with voice telephony and that the local telephone company is the usual provider of this service. Therefore, we adopt a clearer perspective. Telecommunication is the electrical transmission of speech, data, and visual information over long distances (e.g., fax and TV). Computers communicate to one another through telecommunications networks. (Please see later in this chapter for a description of computer networks.)

Today, telecommunications infrastructure is crucial to productivity gains in offices and industries worldwide. Computer advancements have resulted in the creation of new telecommunications. They are technological relatives based on microchips and running programmes (software accounts for 70% of the development cost of telecom switching). Telecommunications devices are evolving at the same rate as computers in terms of price and market performance ratios.

The Evolution of Computers

The modern computer emerged from counting and calculating machines invented in the past. The Abacus, which has been used since 2400 BC, is the earliest known counting machine invented. It consists of movable beads or objects of similar nature strung on a rod or wire. The beads represent digits and are manipulated to perform operations such as counting, addition, subtraction, multiplication, and square root. The Abacus had been used in several places even before the acceptance and use of Arabic numerals. It was used in the Middle East, North Africa and the Nile Valley in ancient times, as well as in Europe and China.

The enormous evolution of the computer over the past 200 years is mainly responsible for the emergence and development of information technology. The invention of the computer occurred incrementally with various scientists and mathematicians through the years building upon the progress made by their predecessors (Awati, 2021).

Joseph Jacquard, Charles Babbage and Ada Lovelace

In the 1800s, the idea of counting or calculating machines that operated mechanically – instead of the manual operation of The Abacus – took hold. During this period, the Jacquard Loom was developed by Joseph Jacquard. The loom used a chain of punch cards fed into it to instruct the system on how to make fabrics with intricate woven patterns.

The operation of the loom, which, with punch cards, directed the system to perform various tasks, is similar to what the computers of today also do. Until electronic devices came into use well into the 20th century, such devices used the Jacquard Loom's system of automatically issuing machine instructions.

The Jacquard Loom was a massive inspiration to Charles Babbage – considered the father of the computer – following his work on the Difference Engine and the Analytical Engine. Babbage machine designs were the first to incorporate many concepts in modern computers, including separating data and programme memory, instruction-based operations, programme control using conditional jumps, and a discrete unit for input/output.

In 1822, Babbage developed the Difference Engine as a mechanical device that would assist humans in calculating polynomial functions automatically. It is therefore seen as the first computing machine. The device prepared numerical tables by using a small set of polynomial coefficients. Up till now, the capabilities of the Difference Engine for computing logarithmic and trigonometric functions are leveraged in computing many of the mathematical calculations carried out in engineering, science and technology and in directing the paths of flights and ships.

In the 1830s, Babbage started working on the Analytical Engine, a successor to the Difference Engine. The engine was proposed as a general-purpose mechanical computer designed to operate on a punch card system.

The design of the Analytical Engine featured a unit for carrying out arithmetic calculations. This control unit enabled the system to loop and detour different system branches, assisted by a memory component. Therefore, the Analytical Engine was very much like a modern computer system by the structure or nature of its components and logic of operation. Unfortunately, Charles Babbage could not fully build the Analytical Engine during his lifetime for a lack of funding.

In 1843, Ada Lovelace met with Babbage and became his life-long associate. She was asked in 1843 to translate a paper written by an Italian Engineer, Luigi Federico Menabrea. She not only did the translation; in addition to this, she also went further to enhance the work of Babbage greatly. She improved Babbage's design of the Analytical Engine and provided notes and instructions on how to operate the machine. Those operational instructions became known as computer programme. This is why Ada Lovelace is rightly referred to as the first person to have written a computer programme. In other words, she was the first Computer Programmer.

In her notes, Ada Lovelace showed how a set of instructions could be given to the Analytical Engine to handle data, numbers, letters, symbols and information. She also described how to programme the machine to loop and repeat the loop as often as possible and how to branch to other areas of the programme. Today's computer programmers do that too, and broadly refer to it as looping.

Today, most of the data processing and execution capabilities and coding language employed in IT, such as conditional branches and loops, arose from the early work of Joseph Jacquard, Charles Babbage, and Ada Lovelace. However, the modern computer can be traced to 1936, when the first calculating machine, considered to be the first programmable computer, was invented by Konrad Zuse.

In the 1940s, during the Second World War, Alan Turing, a British mathematician working alongside other codebreakers for the British government, invented the Bombe. This codebreaking machine intercepted and deciphered encrypted communications from the German Enigma Machine. It is estimated in some quarters that without the invention of the Bombe, the war would have lasted longer, and many more lives would have been lost.

During the same time, John Mauchly created the first programmable general-purpose digital electronic computer, ENIAC

(Electronic Numerical Integrator and Computer), to assist the United States army with ballistics analytics. The machine could calculate thousands of problems per second, enabling the US army to calculate the trajectories of artillery shells efficiently.

The ENIAC weighed 30 tonnes and was housed in a 1,500-square-foot room with 40 cabinets, 6,000 switches and 18,000 vacuum tubes. Besides the fact that ENIAC could perform thousands of calculations per second, it could be easily reprogrammed for different tasks, giving it multiple uses. Since the creation of ENIAC, advancement in technology has led to consistent improvement in the computers that have been made.

When the ENIAC was being improved upon, Alan Turing developed what he called The Turing Test – initially called The Imitation Game – in 1950. The Turing Test is a test that determines the capability of a machine to exhibit intelligent behaviour similar to that of humans. Turing proposed that we can decide if a system has Artificial Intelligence and if that system can process data and information so that any reasonable person can say that the outcome is similar to how humans carry out the same tasks. Thus, the Turing Test has become the cornerstone for determining machine intelligence in computer science.

Between 1956 and 1963, a new generation of computers emerged. They were smaller and used transistors rather than vacuum tubes, as in the past. However, the most impactful computer changes occurred between 1971 and 2000. This was a period in which computers could be provided with millions of transistors in a single chip circuit. This led to the invention of the Intel 1004, the first commercially available microprocessor. The computers were also equipped with a Graphic User Interface or GUI.

The invention of the Intel 1004 microprocessor ushered in the era of personal computers, specifically desktop computers. The arrival of personal computers, with their ability to create networks, ultimately led to the invention of the Internet in the early 1980s. Eventually, high-end microprocessors were developed, registering as many as 19 billion transistors. Soon after, manufacturers incorporated high-end processors in smaller devices used in businesses and homes.

Since then, smaller computers, including laptops, tablets, smartphones, and other hand-held devices, have been created.

Though way smaller than ENIAC, these devices operate at a far higher speed and contain multiple times more memory and storage capacity than ENIAC.

At present, 5th generation computers are being developed to headline the fourth industrial revolution. These computers are designed to harness the progress made in artificial intelligence and incorporate this into the systems. With the incorporation of artificial intelligence (AI), 5th generation computers will be smart machines that mimic cognitive functions. This means that computers will have the capacity to perform tasks that typically require human intelligence; they will be able to think like humans. Systems that include AI capabilities are already being used in numerous real-life applications, while others are still in the developmental stages.

Timelines of Important Milestones in the Evolution of Information Technology

As stated earlier, from 1956 onward, successive generations of computers were developed – smaller in size and in which transistors replaced vacuum tubes. Those generations of computers came with several valuable devices such as screens, text editors, the mouse, hard drives, fibre optics, and integrated circuits. During the same period, government agencies, the military, and universities were most invested in and found more use for the developments that were taking place in the world of computers. However, it was not long before these developments' usefulness spilt over to the corporate world (Complete I.T., 2021). During this period, various applications were being developed for office use and other purposes. For example, Microsoft Office applications like Word, database management systems and spreadsheets.

Following this, various programming languages, including Pascal, FORTRAN, Basic, C, C++, LISP and Prolog, were created to run databases and write and update networking software. This was all geared towards developing systems that would allow the design, creation, adaptation, and maintenance of systems that can be used to assist in various human endeavours.

Today, a lot of the technologies in use in IT owe their existence to the Internet. However, the Internet itself emerged from ARPANET.

This network was developed under the direction of the Advanced Research Projects Agency (ARPA), an arm of the U.S. Defence Department, in the 1960s (Featherly, 2021). The project was conceptualised as a global or intercontinental network of computers by scientists at several universities across the United States of America that undertook research funded by the Department of Defence. The aim was to make it possible for computers at the institutions connected to the network to share information.

The ARPANET was initially comprised of just four computers, but today it has ballooned into a network interconnecting millions of computers and networks. In other words, membership of the ARPANET network was initially limited to only those institutions and organisations that undertook research and contracts for the Department of Defence. Today, however, it has become a global network of interconnected networks to provide a worldwide platform for sending and receiving data and information.

Subsequently, Transmission Control Protocol (TCP) and Internet Protocol (IP) were developed in the 1970s, enabling different kinds of computers on other networks to communicate or share data and information virtually via Packet Switching. TCP/IP enables every electronic device on the planet which is connected to a network to "speak" and exchange any form of data in milliseconds. The invention of TCP/IP led to the use of emails worldwide. Thus, email has become a handy way for businesses and individuals to communicate and exchange messages.

On January 1, 1983, the Internet arose. This resulted from ARPANET and the Defence Data Network of the USA moving on to using the TCP/IP standard. From that day, all networks could now be connected by a universal language – TCP/IP.

When Tim Berners-Lee invented the World Wide Web in 1989, it became possible to access billions of online pages. The World Wide Web allows documents to be connected to other documents by hypertext links, enabling the user to search for information by moving from one document to another.

In the early and mid-1990s, the first search engine, web browsers and Domain Name System (DNS) were developed. In 1998, the search engine giant Google was founded. With the invention of DNS, web browsers, and search engines, users could find information on the World Wide Web and display it in one place.

In 2003, WordPress, an open-source content management system, was created. WordPress is a tool that makes it possible and easy for anyone – even those who are not developers or programmers – to build a website and manage essential aspects of the site.

With the developments that have occurred in the evolution of the computer, the invention of the Internet, the invention of the email, the invention of the World Wide Web, Domain Name System, search engines, web browsers and WordPress, data could be transferred or communicated on the Internet via the World Wide Web from one computer to any other computer across the world.

Computer Networks

A computer network is a system of computers and associated devices that are connected electronically through their communication links/nodes; in such manner the network is able to interact with its different connected devices in the transfer of varying types of data and information, with the aid of various applications.

Usually, the computers that constitute a computer network are interlinked through various wired and wireless connections. The connections that use wires do so with copper and fibre optic wires. Wireless connections, on the other hand, transmit data and information through the assistance of waves with the help of electromagnetic technology.

While data can be transferred from one computer to another using physical media such as DVDs and external hard drives, a computer network allows it to share data between computers without engaging any physical devices.

Types of Computer Networks

There are four significant categories of computer networks. They are to do with the size of the network, the nature of the media used for transmission, the type of network topology used, and the type of method used for managing the network.

Based on the size, there are three categories of computer networks, namely:

1. **Local Area Network (LAN):** This is usually made up of a network of computers within a single site, typically an individual building. A good example is a network of computers inside an office or on one of the floors of a building.
2. **Wide Area Network (WAN):** This is a network of computers extending across a significant geographical location, such as a region or an entire country. This is precisely what the Internet is; it is a massive WAN.
3. **Metropolitan Area Network (MAN):** A MAN falls between a LAN and a WAN. In effect, it is not quite as big as a WAN but is more significant than a LAN. A computer network within a particular city could be described as a Metropolitan Area Network.

Based on the nature of the media used for transmission, there are two categories of computer networks, namely:

1. **Bound/Guided Media Network:** The bound or guided media is usually referred to as wired media. This network type establishes a physical interface between two nodes in the network that are linked together. Within the network, all physical connections point in the same direction. Such networks can make use of twisted pairs, coaxial cables, optical fibre cables, and other types of interconnection. Twisted pair is a common networking medium because it is more lightweight, less expensive, more flexible, simple to install, and offers faster rates than coaxial wires. Although coaxial cables have less crosstalk than twisted pair cables. For long-distance communication, coaxial cables are often utilised. The two most common coaxial cable kinds are RG-6 and RG-59, where RG stands for "radio guide." Light waves are used for transmission in optical fibres. Optical fibres have no problems with crosstalk, EMI, or attenuation. Although optical fibre cable costs less than coaxial and twisted pair cables, the extra optical components required for installation make fibre optic the most expensive of all cables. LAN and MAN can be used for bounding/guided media networks.
2. **Unbound/Unguided Media Network:** Unbounded/unguided media is also known as wireless media. In this form of network, electromagnetic transmission does not need a physical interface.

In such networks, radio waves, microwaves, infrared, and other technologies can be employed for connection. MAN and WAN can be utilised to create an unbound/unguided media network. The ability to extend networks without upsetting present users is one of the benefits of an unguided transmission medium. In contrast to the guided transmission medium, they move at a slower speed.

Based on the type of network topology used, there are two categories of computer networks, namely:

1. **Physical Topology:** The physical topology of a computer network explains how the nodes, or computers, are interconnected. There are some tasks involved in setting up a computer network (links, nodes, etc.), including the placement of devices, the number and types of network adapters, the cable used, and the installation of code. Also, the physical configuration of the network consists of workstations, network nodes, and other network equipment.
2. **Logical Topology:** A logical topology defines how data flows between computers. It is an associated part of a network protocol. It specifies how data is sent through a network and the path it follows. The topology of a logical network specifies how data is conveyed via a physical network structure. Virtual network resources and cloud resources are included in the logical topology.

Based on the type of method used for managing the network, there are two categories of computer networks, namely:

1. **Client-Server Network**: Client-server networks enable a client to execute a programme and have access to data stored on the server. One computer serves as the server and all other computers, known as clients, are in this type of network. In most cases, a client-server network is chosen over a peer-to-peer network that lacks a centralised server for network management. The failure of the computer of one client has no impact on the other clients' computers. A server failure is likely to affect every computer on the network, and overall setup costs are higher than for a peer-to-peer network.
2. **Peer-to-Peer Network:** Peer-to-peer networks allow information to be sent between peers without the use of a centralised server. In other words, each server node serves as both a client and a server.

Thus, there is no expense associated with purchasing a server or hiring a network specialist. Each user sets their file-sharing rights and may choose which files they want to share or keep private. One computer handling several tasks and being accessible by others might cause performance to be slower. Also, as each client computer needs its anti-virus software, security may be more expensive in this network management category.

How the Internet Facilitates Exchange of Data Between Computers

The term "Internet" is the collective name for the worldwide networks that act as one coherent system. The concept of the Internet is similar to the traditional telephone system, whereby a user enters a series of digits or numbers to enable him speak to someone anywhere across the world. However, with the Internet, we are not limited to only voice communication; we can send and receive various forms of data – including text, images, and videos – from any electronic device, so long as it is connected to the Internet (Foundation of the Internet., 2022).

The Internet is configured to connect all computers worldwide via the respective independent networks. Various data run through these networks from one computer or electronic device to another without congestion.

Underwater or submarine cables, laid at the bottom of the sea and connecting land-based telecommunication stations, are the Internet's backbone. These cables are responsible for the interconnectivity of computers and electronic devices across the world. They do so with the aid of fibre optics, through which data and information are captured as light waves, which are then passed on through the help of lasers through thin layers of glass fibres. They carry telecommunication signals across stretches of oceans and seas. Currently, more than 450 submarine cables are traversing each of the continents of the world by sea. According to Brake (2019), Griffiths (2019) and Swinhoe (2021), they handle about 95% of the international voice and data traffic of the world.

Practically, every entity, be it a person or an organisation, communicates data and information today through submarine cables to transmit their data and information across the world. As a matter of fact, anyone accessing the Internet, whether on a mobile phone, tablet, laptop or any other electronic device, does so through the connectivity of submarine cables. A consortium of telecommunications carriers majorly owns submarine cables. However, big Internet enterprises are investing heavily in intercontinental submarine cables.

Due to the connectivity that submarine cables provide between land-based telecommunications stations across the various continents of the world, there is no need for cables to run from every computer to another (Swinhoe, 2021). Instead, computers are linked to operators who connect to the next operator, who connects end users to the final destinations. In the case of the Internet, the operators are Internet Service Providers (ISPs) who provide a physical connection to their network to allow users access to the Internet.

The Internet is profound and highly enabling today. Through it, anybody with a computer or any other electronic gadget, such as a smartphone, can transfer data to any other computer or digital device interlinked by the Internet.

Data Transfer or Communication

As previously said, a computer is a handy tool for collecting, processing, storing, and transferring or communicating data. The evolution in the computer's development and the Internet's invention has greatly facilitated this process of transferring or sharing data.

Data refers to discrete facts. They come in symbols, characters or even quantities that a computer can thus manipulate. When a combination of data is presented with some meaning, they are referred to as information. In other words, data are concrete facts, whilst information relates to combinations of data that are processed in a manner that provides some meaning to us with which we can solve some problems or carry out our decision-making.

Data communication is the process by which various computing technologies – including telecommunications systems, networks of computers, and radio/satellite communication – are used to transfer electronic or analogue data from one computer node to another.

Data could be of various sizes. It is obtained or processed at one point and transferred or communicated to multiple other points through bits and bytes over analogue or digital media – such as copper wires, fibre optic cables, and wireless signals.

When transferring or communicating data across the world, from one point to another or between participating parties through a network of computers, it is always imperative to ensure that the transfer is seamless and without hitches. To achieve this, some Internet protocols are used. The relevant protocols are FTP (File Transfer Protocol), HTTP (Hypertext Transfer Protocol), and HTTPS (Hypertext Transfer Protocol Secure). Through the above protocols, data transfers are seamlessly and optimally carried out across the Internet.

Components of Data Communication

There are five components in a data communication system, as shown in **Figure 4**:

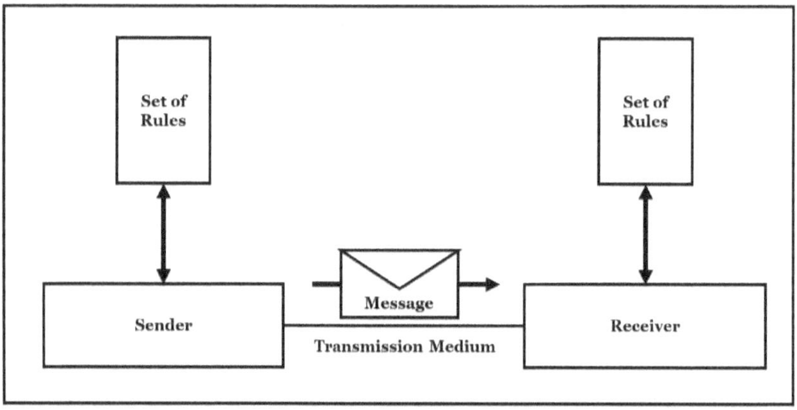

Figure 4: A data communication system
Adapted from https://archive.mu.ac.in/myweb_test/syllFybscit/dcn.pdf

1. **Message**: this refers to the data or information a sender transmits to a recipient.
2. **Sender**: this is the individual/device through which data/information is transmitted to someone else.
3. **Receiver**: the recipient/device to which the sender transmits the information.
4. **Transmission Medium**: this refers to the medium through

which data and information are transmitted from a sender/sending device to a receiver/recipient device. This can be through the use of connected wires or wireless mediums.

5. **Protocol**: this is the transmission rules and standards through which a sender sends, and a receiver receives data being communicated.

Forms of Data Communication

There are various forms in which data is communicated. They include the following:

1. **Text**: Data in text form includes combinations of letters of the alphabet in lower and upper case. In this form, data is stored as patterns of bits encoded in such systems as ASCII and Unicode.
2. **Numbers**: Data in the form of numbers is often captured as a mixture of 0 to 9. Like in text, data in the form of numbers is also stored as patterns of bits encoded in such systems as ASCII and Unicode.
3. **Images**: These data are usually captured in the form of pixels. It refers to the smallest element of images. An image is a matrix of pixel elements represented in bits form. If an image is composed of black, white, or other colours, it needs 100 bits of memory space to store. However, if the image is in more than two colours, say, black, white and grey, it would require 200 bits of memory to be stored. The standard image formats include JPG, PNG, BMP, GIF, etc.
4. **Audio**: Is data captured in sound format and thus can be stored and broadcasted to either one person or many people at once. For example, the sound from a radio set is a form of audio data. The data captured or transmitted in the audio format is transferred continuously instead of as discrete bits.
5. **Video**: This is data broadcasted as a picture. It usually occurs as continuous, analogue signals. The analogue signals must first be converted to a non-continuous, digital format for a computer to process them.

Information Technology Today

Today, Information Technology has developed into a field that extends well beyond the domain of mathematicians. Various scientists and engineers from different backgrounds and skill sets, including programmers or software developers, network engineers, cybersecurity experts, database administrators, system engineers, business analysts, information system managers, and others, form the network of professionals who occupy the information technology ecosystem.

The World Wide Web has tremendously transformed and enormously increased the use of computers to communicate amongst users worldwide. IT, as pointed out earlier, includes hardware systems like mainframes, quantum computers, workstations, personal computers, tablets, smartphones, nano-computers, etc. Cloud computing, invented in the 1960s, is now a significant part of the information technology strategy of many organisations.

Besides cloud computing, various other technological developments have expanded the frontiers of the world of information technology. Some developments include the Internet of things, artificial intelligence, machine learning, robotic process automation, big data, 4G and 5G technologies, and others. These are collectively called exponential technologies. We will look at exponential technologies in Chapter 5, but first, we will take a closer look at the hardware component of computers in Chapter 4.

References

Awati, R. (2021, July 22). *A brief history of the evolution and growth of IT.* TechTarget. https://whatis.techtarget.com/feature/A-brief-history-of-the-evolution-and-growth-of-IT

Brake, D. (2019). Submarine cables: critical infrastructure for global communications. Information Technology & Innovation Foundation, 1-11. Retrieved May 10, 2022, from https://www2.itif.org/2019-submarine-cables.pdf

Brown, T. (2020, May 11). *What Is Information Technology? A Story of Evolution.* IT Chronicles. https://itchronicles.com/it/what-is-information-technology-a-story-of-evolution/

Castagna, R., & Bigelow, S. J. (2021, August 5). *What is Information Technology? Definition and Examples.* TechTarget. https://searchdatacenter.techtarget.com/definition/IT

Complete I.T. (2021, January 13). *The History of Information Technology.* Complete IT. https://www.complete-it.co.uk/the-history-of-information-technology/

Featherly, K. (2021, March 23). *ARPANET*. Encyclopaedia Britannica. Retrieved May 10, 2022, from https://www.britannica.com/topic/ARPANET

Foundation of the Internet. (2022, April 7). Encyclopaedia Britannica; Encyclopaedia Britannica. https://www.britannica.com/technology/Internet/Foundation-of-the-Internet

Griffiths, J. (2019, July 26). *The global Internet is powered by vast undersea cables. But they're vulnerable.* CNN. https://edition.cnn.com/2019/07/25/asia/Internet-undersea-cables-intl-hnk/index.html

Leavitt, H. J. & Whisler, T. L. (1958). Management in the 1980's. Harvard Business Review, 36 (6), 41-48.

Chapter 4

Hardware Development in Africa

As it was pointed out in Chapter 3, the hardware of a computer is the physical parts of the computer. These include a central processing unit (CPU), motherboard, hard disk drive, monitor, random access memory (RAM), keyboards, graphics card, and others. A computer consists of hardware and software.

Software is the set of instructions or programmes that tell the hardware what to do and when to do it, such as an operating system, word processing software, Internet browser, and others. The hardware and software collaborate to make a computer system work.

Although the software design differs for various computers, such as desktops, laptops and iPad, due to their different sizes, they all possess the same core hardware. Without hardware, there would be no way for a computer system to run the software.

Hardware for computers comes in different systems, not just personal computers. Instead, it refers to all computer systems, including supercomputers, mainframes, minicomputers, workstations, laptops, smartphones, etc. The hardware is embedded in many devices, including cars, microwaves, and CD players.

When tech companies in the developed world build computer hardware, they make them suit their needs and align with their business objectives. In like manner, African countries need to concentrate on developing their hardware to meet local needs and compete with the developed world.

When the world underwent transformations as a result of the first, second and third Industrial Revolutions Africa lagged behind.

Consequently, Africa has depended on the industrialised world for most of its products, including products of technology.

Despite these challenges, Africa has been able to do well regarding mobile phone usage and Internet access coverage. In 2016, the International Telecommunication Union (ITU) reported that about 80.8% of Africans have mobile phones (Mushakavanhu, 2017). This rapid transformation has occurred mainly because engineers on the continent have always found ways to adapt the technology that powers mobile technology to suit the specific needs and dynamics of the continent.

The hardware industry in Africa is largely untapped, with various factors constraining its development, including poor electricity and high startup funding. Nevertheless, African countries should focus on developing the hardware sector on the continent. Self-sufficiency and exportation of hardware products remain essential for the expansion of Africa's economy. As a result, a significant number of hardware manufacturing companies are required.

Information technology is a vital part of economic growth in Africa. Within the broad sets of technologies that make IT, computer hardware is essential, since it is impossible to run a software or use a computer to execute any task without hardware. Both private investors and the government of respective African countries will do well to support African hardware startups to ramp up the production of computer hardware on the continent

External and internal computer hardware

Computer hardware is classified into internal and external components. External hardware is generally attached to a computer to boost its performance or add-on features, while internal hardware is necessary to run properly.

Internal hardware components

The computer's internal hardware components are responsible for operating system (OS) instructions. These include the following:

- **Motherboard:** This is the computer's main panel where all critical components like random access memory (RAM), firmware, and internal and external buses interface are connected. The printed circuit board (PCB) makes up the complex electronic system and houses the central processing unit (CPU) and several internal hardware. See **Figure 5.**

Figure 5: Photograph of a motherboard

- **CPU:** The CPU is the part of the computer that carries out digital instructions and is responsible for processing data. Although CPU's design, configuration, and implementation have evolved significantly since the first models, the basic operating principle of CPUs has remained the same. Nearly all CPUs follow four basic steps: *fetch*, *decode*, *execute*, and *writeback*. See **Figure 6** below.
  The first step "fetch" starts by getting instructions from memory. In the decode step, the instruction is parsed and interpreted more systemically due to its relationship to other CPU components. After completing the fetch and decode actions, the execution step is performed. In this step, various portions of the CPU work together to achieve the desired operation. The final step "writeback" writes back the query results to memory.

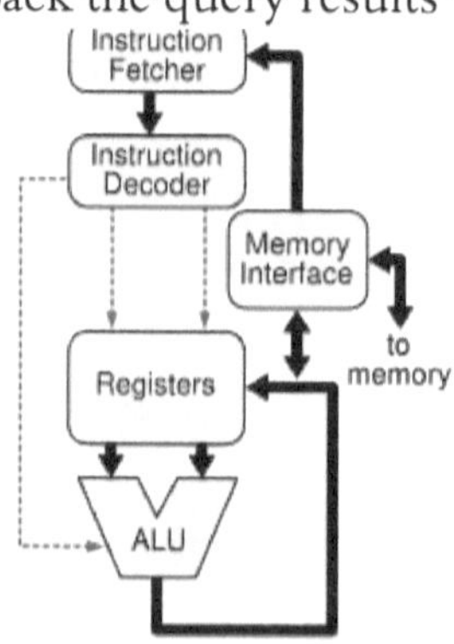

Figure 6: Instruction Cycle of CPU

- **RAM:** RAM, or dynamic RAM, is a type of memory that temporarily stores data and allows computer programmes to retrieve it in any order without altering the data. When a computer is turned off, the RAM is immediately cleared. Meanwhile, distinctions between these different types of RAM include: writable versus read-only, static versus dynamic, volatile versus non-volatile, etc.

Figure 7: Random Access Memory

- **Hard Drive:** Hard disk drives (HDDs) store data in various formats, including programmes, operating systems, and device files. All computers ought to have a storage device, and HDDs are just one instance of a kind of storage device. HDDs are usually installed internally in computers, consumer electronics technologies and enterprise disk arrays in data centres. They can store files using magnetic disks. In addition, hard disk drives oversee how data is accessed from and written to the hard drive installed inside the computer.

Figure 8: Internals of a 2.5-inch hard disk drive

- **Solid-state drive (SSD):** The leading alternative to hard disk drives is solid-state drives. NAND flash memory technology is used in solid-state storage devices to facilitate data storage even when the computer is off. Even though SSDs do not use magnetic fields to store data, the read performance remains stable no matter where the data is located. The main reason SSDs are becoming a preferred storage choice is because they are more reliable than HDDs.

- **Optical Drive:** This is a type of storage disk drive that reads and writes data with the help of laser beam technology. Optical media, such as CDs and DVDs, are commonly used to store data. At first, optical discs seemed like a suitable replacement for hard drives, but they've since lost ground to flash-based SSDs in capacity and performance. Although optical discs are more durable and less prone to damage than tapes, they are still slower than standard hard drives. It offers lower storage capacities than either. Over time, optical media have taken different forms, including LaserDisc, HD-DVD, write-once, read-many (WORM) optical cartridges, etc. It is clear from the aforementioned that three formats of optical drives have appeared as standards in today's optical storage business:

Figure 9: Optical disk formats

Adapted from https://www.techtarget.com/searchstorage/definition/optical-disc

i. Compact disk (CD). The CD illustrates how optical storage technology has evolved. At first, CDs were only known as pre-recorded, read-only disks; soon, they became known as recordable discs and rewritable discs that could be used for data storage.

ii. Digital versatile disc (DVD). The DVD was initially designed to be a read-only storage medium, just like CDs, but it was able to store more data. DVDs can store a large amount of data, up to 4.7 GB on a single-layer disc and 8.5 GB on a dual-layer disc. With the introduction of DVDs, people started saving their data on recordable and rewritable discs after realising the data storage possibilities.

iii. Blu-ray. The Blu-ray disk appeared as a clear leader in today's optical storage business. Blu-ray disks use a blue laser to read and write data, which makes streaming video quality better than CDs or DVDs. With Blu-ray disks capable of storing so much data, it's no wonder they're often used for storing pre-recorded feature films and can store up to 128 GB of data.

- **Heat sink:** This passive piece of hardware removes heat from components to reduce their temperature to guarantee they function well. The heat sink is designed to increase the device's surface area in contact with the cooling medium. The air velocity, the material used, and the protrusion on the heat sink all affect its performance. In computers, the heat sinks are placed directly on top of the CPU, which generates the most heat. Heat sinks keep CPUs, GPUs, certain chipsets, and other components cool.

Figure 10: Heat Sink

Source: Image by fir0002flagstaffotos@gmail.com Canon 20D + Canon 70-200mm f/2.8 L - flagstaffotos.com.auOwn work, GFDL 1.2, https://commons.wikimedia.org/w/index.php?curid=1403949

- **Graphics Processing Unit (GPU):** This is a chip or electronic circuit developed to accelerate graphics rendering for display on an electronic device through quick mathematical calculations. The chip-based device processes graphical data simultaneously and often acts as an extension to the main CPU, making them useful for machine learning, video editing, and gaming applications. Although GPUs are most commonly associated with gaming, they're also used in various applications.

- **Graphics card:** A type of hardware known as a graphics card supports the display of graphics on a device. It includes a powerful graphics processor and fast memory to help with graphics processing. The graphics card has a built-in cooling system to keep the heat away from the GPU. A graphics card allows you to display 3-D images, rasterise images, and get high pixel density. A graphics card also comes with various interfaces, including accelerated graphics port (AGP), high definition multimedia interface (HDMI), a TV connection, and support for multiple displays. Graphics cards can be integrated into the motherboard or attached via an extension card.
- **Network interface card (NIC):** A Network Interface Card, or NIC, is a small electronic circuit board or chip that helps connect your computer to a network. Almost all new computers have Ethernet capabilities built-in. However, a computer can have multiple NICs, including wired and wireless. A wired NIC usually requires a connection to a network with an Ethernet cable or fibre optic cable. Wireless NIC adapters use radio signals commonly used on networks to connect to the Internet. All Wi-Fi devices broadcast and receive radio-frequency waves using the transmitter, which is often not visible. Computers, smartphones, and tablets are examples of devices that can access the Internet using Wi-Fi because they come with an inbuilt Wi-Fi NIC. Printers can have Wi-Fi built in for connectivity.

Other components that may be found in a computing system, such as USB ports and power supplies, are also pieces of internal hardware.

External hardware components

External hardware items are attached on the exterior of the computer. These devices are created to supply instructions or results to the software.

Standard input hardware components include the following:

- **Mouse:** A mouse is a small, hand-held device that allows users to interact with objects on a computer screen. It can be either wired or wireless.
- **Keyboard:** A keyboard is an input device that has a typical QWERTY key layout. It lets people type in texts, numbers, or special characters.
- **Microphone:** Microphones are devices that transform sound waves into electrical impulses for computer-based audio communications.
- **Camera:** A camera takes pictures and sends them to a computer or over the Internet.
- **Touchpad:** To move the pointer on a display screen, a touchpad is an input device that can be either external or incorporated into a computer. Usually, it serves as a replacement for an external mouse.
- **USB flash drive:** A USB flash drive is a portable, detachable storage device that employs flash memory and can be connected to a computer.
- **Memory card:** A memory card is a portable storage device that can store pictures, music, or videos.

Other input devices include joysticks, styluses, and scanners. Output hardware components typically include the following:

- **Monitor:** A monitor is an electronic output device – a screen–that shows what's going on inside a computer.
- **Printer:** Printers turn electronic data into text and images that can be printed.
- **Speaker:** A speaker is an electronic device that emits sound from an external source.
- **Headphones, earphones, ear buds:** Similar to speakers, these devices produce audio only audible to the person using them.

Hardware Development in Africa

Hardware development is a complex stage incorporating the invention of electronics and mechanical system components where diverse computational processes are executed and optimised. The relationship between these components is realised through a hardware interface.

In order to industrialise a society, hardware is essential. It represents the mechanics of the physical economy. Hardware development is necessary to meet Africa's needs for goods and services, improve the country's infrastructure, provide employment, and reduce poverty. Today, leaders acknowledge that hardware innovation is critical in helping African countries attain their goals of successfully advancing to the next stage of economic development. In its Agenda 2063, the African Union prioritised the manufacturing industry.

African governments are working hard to attract investment by implementing new strategies, including developing infrastructure, increasing regional integration, and creating special economic zones for businesses in essential sectors. These developments will drive demand in several sectors, including hardware, food and beverages, pharmaceuticals, financial services, healthcare, and education.

The computer hardware market grew by 7.6% from $1,129.39 billion in 2021 to $1,215.76 billion in 2022. The computer hardware market is predicted to reach $1,568.25 billion in 2026 at a CAGR of 6.6% (The Business Research Company, 2022). Exploiting hardware innovation could drive national development.

Also, these trends create opportunities for countries like South Africa, Egypt, Nigeria, and all regional outperformers in the Global Manufacturing Competitiveness Index. Countries such as Ethiopia, Morocco, Rwanda, and others have recently started to allow manufacturing and industrial development. Supporting local hardware industries can lead to many positive outcomes, especially if African countries want to adapt hardware technology to meet local needs.

An advantage of hardware innovation is that it galvanises large swaths of employees and positions them in productive and best-paying jobs. After years of implementation, this process has led to developing prosperous nations such as the United States, United Kingdom, France, Japan, and Germany. Recently, China has seen a rise in industrialisation, which has helped propel it into one of the world's fastest-growing economies. These examples show how industrialisation can cause rapid structural change, enable development, and alleviate poverty and unemployment. However, this narrative excludes many African countries because they cannot produce their hardware. A government that wants a better performance in development would invest more in hardware projects.

In 2019, African countries imported machinery and electronics from China worth $38.84 million. The value of imported goods in 2013 was 167.15 million U.S. dollars (Saleh, 2021). Additionally, business-to-business spending in manufacturing in Africa is projected to be over $1 trillion by 2030 (Signé, 2018). The African hardware industry needs more hardware manufacturers and startup hardware developers.

Although hardware development has high entry barriers because it is highly capital-intensive and labour-intensive, with a high rate of products liable to obsolescence (at least for certain types of hardware), it has an equally higher economy of scale. The favourable factors in Africa for hardware production, combined with the low cost of African labour, creates a clear opportunity.

Hardware companies are springing up in Africa

The continent may not have the biggest Original Equipment Manufacturers (OEMs) in the world like Apple, IBM, Hewlett Packard (HP), Ericsson, Philips Electronics, Toshiba, The Boeing Company, GE Medical Systems, Volkswagen, and Toyota. Still, a few African countries have begun the journey to becoming self-sufficient in manufacturing hardware and other technologies by creating appropriate policies and beginner efforts. The significance of hardware manufacturers in any economy cannot be overestimated.

These OEMs determine how much the economy would spend on imports and how much it would make on exports. The future of Africa's economy is linked to the success of its hardware businesses.

Here, we examine the emerging African hardware manufacturing firms:

1. Semiconductor Technologies Limited (STL) in Kenya

Semiconductors are tiny pieces of silicon that control the operation of electronic equipment. They ensure our computers, phones, planes, and other devices function properly to do our jobs effectively.

There is much demand for semiconductors, which is why businesses can make more advanced electronics. The global semiconductor industry is dominated by a few large companies from the United States, Taiwan, South Korea, China, and the

European Union. In 2021, a semiconductor fab was established by Semiconductor Technologies Limited (STL) in Kenya. The company set the pace for local production of integrated circuits, sensors and related nanotechnology products. The firm produces a wide range of products, including mobile phones and television sets, which will be sold locally and internationally (Wanjala, 2021).

2. Global Platinum Group Metals (PGMs) in South Africa

South Africa is the leading manufacturing hub in Africa. Its manufacturing capabilities, efficient logistics network and preferential regional market access position the country as an ideal location for electronics companies seeking to supply their products to Africa. According to Stats SA (2018), the country produces R90 billion worth of different electrical tools and equipment each year. It has a lot of the global platinum group metals (PGMs) supply. PGMs play a significant role in developing emerging fuel cell technology. South Africa's plentiful reserves of precious metals ensure the country has access to various PGMs.

African hardware startups are industrialising

There's an emerging African industrial revolution as startups produce more hardware products. Within a decade, we estimate that African startups have the potential to double their production to nearly $800 billion. About one-third of the growth in the economy will likely come from new hardware startup innovations that substitute imports and meet growing local demand. Without specialised hardware, we can't use exponential technologies in contexts like digital payments and precision farming. Africa's hardware advancements will allow businesses to shift production away from China, making the continent a more cost-effective location for manufacturing.

Hardware startups are popping up all over Africa to help revive the fortunes of the hardware sub-sector of the African tech ecosystem. In 2017, Jon Kornik founded Plentify in South Africa. Plentify integrates the Internet of Things (IoT), artificial intelligence, and a mobile app to create a thermal battery from a water heater using grid-connected technology. The power generated is stored for future use. By doing this, the demand for generators is reduced, and water heating charges are reduced by roughly 50%. The same year, Cape Town-based

Sentian developed innovative security technology products called Sentian i3. The product is an intelligent hub that connects to pre-installed burglar alarm systems, CCTV cameras, home automation systems, and other gadgets over the Internet. This makes them more intelligent and easier to control. As soon as the Sentian i3 detects which motion sensors have been triggered, a video verification clip is immediately delivered to the user's smartphone.

In 2016, a Kenyan company developed a soil analysis kit to help farmers better assess the quality of their land. The hardware solution includes an electronic sensor placed in the ground to guide farmers' real-time soil conditions. The startup uses big data analytics to improve its understanding of farming trends and produce outcomes. Ujuzi Kilimo adapts and enhances farmer recommendations using the data. In the same year, Ntsako Mgiba, a mechatronics engineer, and Ntando Shezi, a business science graduate, founded the security firm Jonga in Cape Town, a low-cost home monitoring security system consisting of an IoT motion sensor device and a smartphone app. Jonga has mentioned that they have created and tested their working prototypes in Khayelitsha, Cape Town. It is difficult to say if those devices effectively reduced burglary rates.

Dext Technologies, founded in Ghana in 2016, produced a hardware product called Science Set. This science set replicates a real science lab and is perfect for home or school use. It includes cells, resistors, and electrical conductors for science experiments. Solar Freeze in Kenya developed mobile cold storage units powered by renewable energy to help rural smallholder farmers reduce the considerable challenge of post-harvest loss.

e-Maji was launched during the 2014 Southern African AMPION accelerator; the team from e-Maji won first prize at the competition's pitch event with their innovative hardware solution for detecting and identifying chemical water contamination.

Back in 2013, BRCK, a Silicon Hills team in Kenya, designed and manufactured a mobile Wi-Fi hotspot called the PicoBRCK. Chimera IoT, a Kenyan startup, developed intelligent home applications and systems that let you control simple functions like lighting, temperature control, and audiovisual integration. This means you can adapt your environment to your moods and needs. Lumkani developed an early-warning fire detection system specifically for informal settlements. Lumkani's product is internationally recognised for its positive impact on the world (Ventureburn, 2020).

Some hardware startups choose not to have smokestacks, which gives them an edge over competitors. Additionally, this decision makes these companies more compatible and accessible to customers.

The Need for Hardware Industry in Africa

Africa has the potential to be a model for how to use exponential technologies in specific contexts to address pressing challenges.

By 2030, the sectors producing the most value in the continent will be food and beverages ($740 billion), education and transportation ($397 billion), and housing ($390 billion) (Signé, 2018). The GDP will grow by $370 billion in consumer goods, $260 billion in hospitality and recreation, $175 billion in health care, $85 billion in financial services, and $65 billion in telecommunications (Signé, 2018). There will be a lot of growth in African business-to-business spending between now and 2030. The economy's largest sector will be agriculture, agricultural processing, manufacturing, construction, utilities, transportation, wholesale and retail, resources, banking and insurance, and telecommunications and ICT. There is a growing demand for reliable hardware to power economic growth in Africa.

Here, let's briefly dive into those sectors that demand hardware development and suggest steps African investors and leaders can take to translate that need into profitable, sustainable enterprises.

1. **Digital payments:** Every year, MSMEs transact over $19 trillion in cash payments globally (Hattingh et al., 2017). The chance to digitise these transactions and make payments is quite momentous. There are about one hundred million small businesses in Africa, with forty-two million in Nigeria alone. It is estimated that by 2025, a total of $2.1trillion in consumer payments will be made on the continent (Akpata, 2021). The COVID-19 pandemic caused people to use digital payments across Africa. For instance, the Nigerian Inter-Bank Settlement System announced that electronic payments in the country grew 88% year-on-year in Q1 2021 (Techpoint Africa, 2021). The transactions summed up to $155 billion. In the first quarter of 2021, there were 1.48

trillion POS transactions, up 48% over the same period in 2020, showing that more customers are using digital channels for their transactions. Financial institutions and regulators want to digitise transactions because it costs countries much money to print, circulate and replace banknotes yearly (Assoko, 2020). Yet, Africa depends on western countries to provide the digital infrastructure for our cashless policies to work, which is expensive to acquire and often unstable. Africa needs to develop hardware that can be used for digital payments, with features tailored to local needs and also provide value-added services. This will be an opportunity for hardware developers, investors, and startups to focus on designing services for a specific part of the payments value chain that is yet to be addressed.

2. **Precision farming:** Africa has a long history of being a land of opportunity, with plenty of resources. Despite this, Africa has had difficulty translating these resources into a model, resulting in equitable economic and sustainable development. The innovations would have led to a significant increase in business growth in Africa, but we didn't take the necessary precautions to ensure its success.

   The high cost of equipment used by farmers in African countries and the lack of farmers' awareness about precision agriculture are the main reasons why the precision agriculture market is not growing as quickly as it could. There are urgent demands for hardware development of auto-guidance equipment, variable-rate technology, proximate sensors technology, grid sampling, remote sensors, relative sensors, precision irrigation systems, and global positioning systems to meet the growing global demand for farming precision tools. These, in return, will provide local food security, give insights into their crops, improve their yields, and enhance their sustainability.

   The market for precision agriculture is set to continue growing as sustainable farming becomes more popular. The increase in demand for new agricultural practices is due to the growth of the population and the increased need for food.

   The Precision Agriculture Market in the Middle East and Africa is predicted to grow by 11.7% over the next five years, reaching a value of $1.0 billion by the end of the decade (Market Data Forecast, 2021). Precision farming is a system that uses information, technology, and management to increase production efficiency and improve product quality (Agri Farming, 2021).

3. **Digital Healthcare:** In 2017, the size of the medical device market in Africa was $4.9 billion (Stewart, 2018). Market Research Future (2022) forecasted that the African medical devices market will be $7,069.61 million at an approximate compound annual growth rate (CAGR) of 6.3% by 2027. Meanwhile, more than 90% is supplied by imports, which denies Africa the jobs that could be created locally. This will become an issue as the demand for medical hardware continues to grow. The growing awareness and spread of information technology demand for early detection and non-invasive therapies and the development of user-friendly devices are the key drivers of these market expansions in Africa. IFC partnered with hardware startups, manufacturers and financial institutions to establish the Africa Medical Equipment Facility, which provides high-quality medical equipment to the continent. This opportunity will help hundreds of medical hardware developers in seven African countries - Cameroon, Côte d'Ivoire, Kenya, Rwanda, Senegal, Tanzania and Uganda - secure loans to develop medical hardware devices, including a broad range of products such as tongue depressors, thermometers, blood sugar meters, artificial hearts, stents, Sonogram machines, X-ray machines, stethoscopes, and wheelchairs (IFC, 2022).

Remember that hardware innovation is more than ensuring your devices can connect to the Internet. It's about improving the technology itself. Improving hardware innovation has many positive spin-offs, including changing the course of history.

The Need for Government Action

The role of government in promoting hardware development in Africa is important because it creates jobs and helps to reduce poverty.

1. African governments need to re-evaluate their policies and examine how physical, financial, and intellectual assets are handled. The more precisely things are understood, the easier it will be for the economy to function and for hardware development in Africa to

be successful.

2. The significance of the role of African governments in education is important herein producing the minds with essential skills for hardware development. This is because poor education is the main obstacle to hardware innovation. Hardware developers need to be selective when choosing technology, as they will need to account for cultural differences. All of this calls for good education. In addition to helping to encourage industrial research and technological change during the early stages of hardware development, the governments can also assist by setting up bricolage centres. Ignorance of technology can hinder a country's ability to progress technologically. China, Hong Kong, Korea, and Singapore all have economies that are doing well now because they boosted their technology sector.
3. Interventions by African governments are frequently needed for technological developments. Because the outcome of the invention is sometimes debatable and technologies might be pirated or stolen, hardware developers typically run considerable risks. These demonstrate why government action is necessary. Thus industries may expend less technological measures than desirable if they cannot get the return on investment expected. These problems can be addressed in a variety of ways:
   a. permitting businesses to patent their products;
   b. providing financial assistance for technological advancements;
   c. encouraging agencies with expertise in R&D to help promote technological development;
   d. establishing bricolage centres that provide all the enabling raw materials, expertise and tools which people with ideas can experiment with, tinker and create new technologies.
4. African governments must provide the necessary physical infrastructure for the industries to function. Despite some benefits to operating parts of large public systems in the private sector, most African countries rely on government provision of these systems.
5. African governments must ensure that hardware developers have access to up-to-date information. Hence, there is a need for government to establish large databases and knowledge-bases with big data. The government can store and use the information to

forecast the technology markets, as well as individuals and private sector organisations.

6. African governments should intervene in the market and help improve the economy by building appropriate infrastructures. America's IT industry was founded on government investments during the 1940s, 1950s, and 1960s. America has been a responsible player in the international community, having helped build the country's industrial infrastructure.

In conclusion, if African countries want to remain successful in the tech industry, they'll need to keep up with the best practices of developed countries, which include involvement and mastery of hardware research and development. They should try to be flexible and revolutionary in their interventions to be most effective.

References

Agri Farming. (2021, July 2). *Precision farming tools - Guide for beginners.* Agri Farming. https://www.agrifarming.in/precision-farming-tools-guide-for-beginners

Akpata, O. (2021, June 4). *Who will capture the merchant payment opportunity In Africa?* Forbes. https://www.forbes.com/sites/columbiabusinessschool/2021/06/04/who-will-capture-the-merchant-payment-opportunity-in-africa/?sh=52fa1e6a2b98

Assoko, J. T.-L. (2020, June 5). *Mastercard: In Africa, our biggest competitor is cash.* The Africa Report. https://www.theafricareport.com/29360/mastercard-in-africa-our-biggest-competitor-is-cash/

Department of Mineral Resources and Energy (2021). Global platinum group metals. South African Department of Mineral Resources and Energy. Retrieved August 27, 2022, from https://www.dmr.gov.za/

Hattingh, D., Leke, A., & Russo, B. (2017, October 2). *Lions (still) on the move: Growth in Africa's consumer sector.* Mckinsey & Company. https://www.mckinsey.com/industries/consumer-packaged-goods/our-insights/lions-still-on-the-move-growth-in-africas-consumer-sector

IFC. (2022). *IFC's Africa medical equipment Facility.* International Finance Corporation (World Bank Group). https://www.ifc.org/wps/wcm/connect/industry_ext_content/ifc_external_corporate_site/health/ifc+africa+medical+equipment+facility

Market Data Forecast. (2021, January). *Middle East and Africa precision agriculture market by type and size-industry report 2027.* Market Data Forecast.

https://www.marketdataforecast.com/market-reports/middle-east-and-africa-precision-agriculture-market

Market Research Future. (2022, August). *Africa medical devices market report - Forecast 2027.* Market Research Future - Industry Analysis Report, Business Consulting and Research. https://www.marketresearchfuture.com/reports/africa-medical-devices-market-2845

Mushakavanhu, T. (2017, July 7). *African manufacturing its own hardware: BRCK Internet in a box [Haller Prize 3rd place].* African Arguments. https://africanarguments.org/2017/01/african-manufacturing-its-own-hardware-brck-Internet-in-a-box-haller-prize-3rd-place/

Saleh, M. (2021, August 27). *Africa: machinery and electronic import from China.* Statista. https://www.statista.com/statistics/1259884/import-value-of-machinery-and-electronics-from-china-into-sub-saharan-africa/

Signé, L. (2018, March 14). *Capturing Africa's high returns.* Brookings. https://www.brookings.edu/opinions/capturing-africas-high-returns/

Statistics South Africa (2018). Quarterly Financial Statistics. Retrieved August27, 2022, from www.statssa.gov.za

Stewart, C. (2018, October 2). *Medical device market size Africa 2018.* Statista. https://www.statista.com/statistics/890876/medical-device-market-size-africa/

Techpoint Africa. (2021, April 20). *Nigeria's $155b e-transactions, ride-hailing drivers strike Andela's global expansion.* Techpoint Africa. https://techpoint.africa/2021/04/20/techpoint-digest-63

The Business Research Company. (2022, February). *Computer hardware market size 2022 and growth analysis.* Business Research Company. Retrieved August 10, 2022, from https://www.thebusinessresearchcompany.com/report/computer-hardware-global-market-report

Ventureburn. (2020, October 2). *SA SME wins a prestigious humanitarian award and € 1 million.* Ventureburn. https://ventureburn.com/2020/10/sa-sme-wins-prestigious-humanitarian-award-and-e1-million/

Wanjala, T. (2021, May 30). *Kenya first in Africa to make electronic chips amid questions over relevance of tech.* Zenger News. https://www.zenger.news/2021/05/30/kenya-first-in-africa-to-make-electronic-chips-amid-questions-over-relevance-of-tech/

Chapter 5

Exponential Technologies

Technology has existed for a very long time and has aided the evolution and development of humans throughout history. From the pre-mechanical age (3000 BC – 1450 AD) to the mechanical age (1450 – 1840), to the electromechanical age (1840 – 1940), to the present era – the electronic age (1940 – to date), technology has continually evolved with each innovation laying a foundation for the next technological improvement.

During the pre-mechanical period, humans communicated broadly through rudimentary pictorial drawings known as petroglyphs. They drew these pictures on rocks. This language was employed to tell stories, keep track of the number of animals one possessed, and indicate one's territory. Finally, writing systems were developed. Apart from Hieroglyphics, another more widely used writing system invented around 3,000 BC, and used widely in the Middle East in ancient times, was known as cuneiform by the Mesopotamians. Thus, it was one of the first writing systems ever developed. The first material for writing with a pen-like object was wet clay. Writing on bark, leaves, and leather, utilising the papyrus plant to make paper, manufacturing paper out of rags, and finally developing modern-day paper became functional writing materials. These early technologies, however, created a new problem as more people used them. This is because the information they captured could not be kept secure for long. As a result, clay tablets and scrolls were created, which eventually led to books and libraries. During this period, numerical systems and the Abacus, the earliest calculator, were also invented.

In the mechanical era, humans made significant technological advances with the invention of the typewriter, spectroscope, the Pascaline and the slide ruler (an analogue computer used for multiplying and dividing). The electro-mechanical era ushered in telecommunications as we know it today (Southern Illinois University Carbondale., 2016). Many technologies that formed the foundation of telecommunication and modern-day telecommunications were invented during this period. These include the telephone, telegraph, Morse Code, and radio. The electronic age constitutes the present era. It started in 1940 when electronic devices with vacuum tubes were invented. ENIAC started the digital revolution, as it is deemed the first computer to be programmed, which has mushroomed today to different programmes running numerous computer systems in the world, whether it is a supercomputer, mainframe computer, personal computer, laptop, or smartphone.

Since the invention of the ENIAC, much technological advancement has led to various innovations that have digitalised the world and accelerated the development of information technology (Hagel, Brown, Samoylova & Liu, 2013). As pointed out in Chapter 3, these innovations include the personal computer, the Internet, World Wide Web, wireless networks, search engines, tablets, and smartphones. This rapid digitalisation has made it easier for a large amount of text, audio, and visual data to be generated, shared and consumed faster.

Over the years, since the beginning of the electronic era, the capacity of the three core building blocks of information technology – computing power, data storage, and bandwidth – has continually increased while the cost of acquiring them has progressively fallen. The graphs below show the relationship between the progressive increases in capacity relative to the continuous decrease in the cost of the three core building blocks of information technology from 1992 - 2020:

From 1992 to 2020, **Figure 11** shows that the cost of computing power decreased from $222 per million transistors to $0.01 per million transistors.

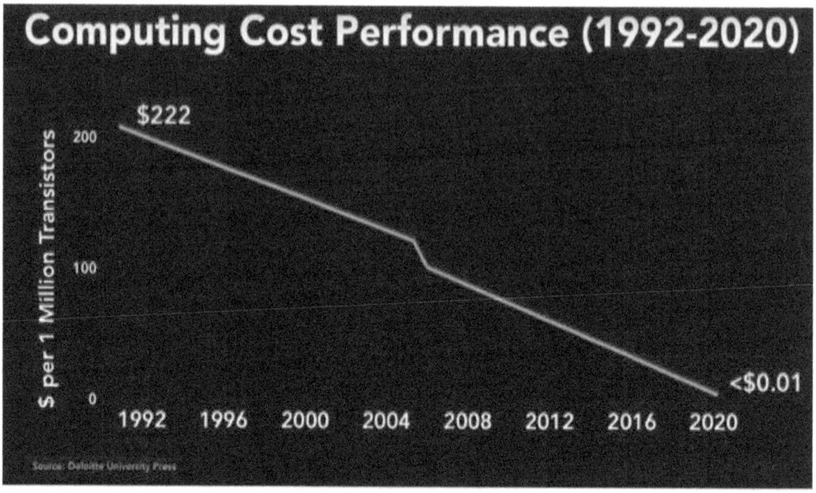

Figure 11: Computing cost-performance (1992 -2020)
Source: Deloitte University Press. DUPress.com

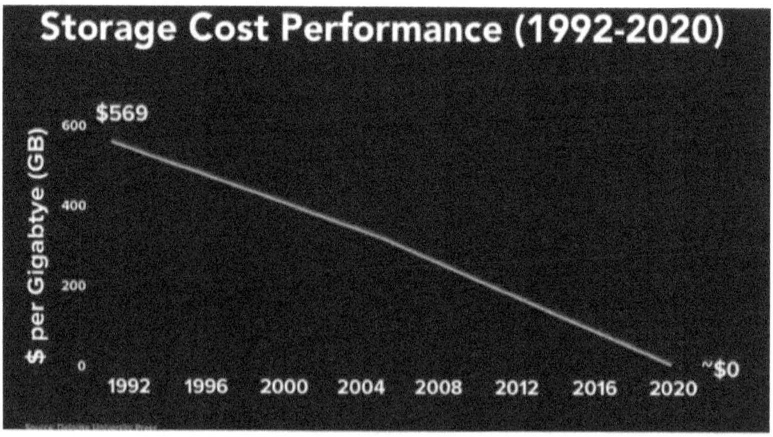

Figure 12: Storage Cost Performance (1992 -2020)
Source: Deloitte University Press. DUPress.com

From 1992 to 2020, **Figure 12** shows that the cost of data storage decreased from $569 per gigabyte of storage to $0.03 per gigabyte.

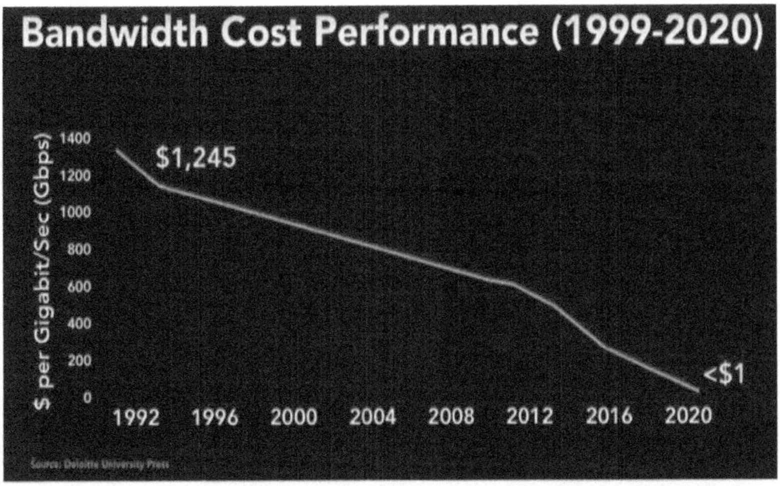

Figure 13: Bandwidth Cost Performance (1992 -2020)
Source: Deloitte University Press. DUPress.com

From 1992 to 2020, **Figure 13** shows that the cost of Internet bandwidth decreased from $1,245 per 1000 megabytes per second (Mbps) to $23 per 1000 Mbps.

The decrease in the cost of transistors relative to their number in the computer has significantly increased the computing power of computers since the invention of the ENIAC.

The decreased gigabytes cost has created much more storage power and richer digital information.

The decrease in the cost of Internet bandwidth makes it possible for data to be collected and transferred faster. The lower price of bandwidth also facilitates richer connections and digital interactions.

These developments have led to numerous innovations in the world of information technology, which have given rise to innovations generally referred to as exponential technologies (Hagel et al., 2013).

Exponential technologies are technologies that change at an extremely rapid speed. At the moment, exponential technologies are the most advanced stage of the evolution of technology. They include the following: artificial intelligence (AI), blockchain technology, Internet of Things (IoT), robots, augmented reality and virtual reality, big data, cybersecurity, 3D and 4D printing, additive manufacturing, 5G and 6G, quantum computing, cloud computing, drones, smart cities, bionauts, and driverless vehicles.

1. Artificial Intelligence (AI)

Artificial intelligence is an aspect of computer science that focuses on designing computers that can mimic the cognitive ability of human beings. Such computers exhibit human intelligence characteristics, such as learning, reasoning, perception, language use, and problem-solving (Woherem, 1993). With these characteristics, such computers can perform tasks that could be carried out only with human intelligence.

The name and concept of "artificial intelligence" were first coined in 1956 by John McCarthy. However, the subject had been developing since the 1930s; in his work entitled *As We May Think*, published in 1945, Vannevar Bush proposed a system that amplifies individuals' knowledge and understanding (Popova, 2012). In 1950, Alan Turing wrote a paper suggesting that machines could simulate human intelligence and perform such tasks as required by human cognitive ability.

In his paper, "Computing Machinery and Intelligence," Turing surmised that in the next fifty years, computers would be programmed in such a manner that they could play what he calls an "Imitation Game," whereby many humans that are interacting or listening to the computer, especially if placed in another room, will think they are interacting or listening to a human being. Therefore, he believed that asking whether computers can think is a mute question, as he feels such programmes could be written. Computers would consequently be able to pass the Turing Test (Wikipedia contributors, 2021).

AI has made gigantic strides since Turing made the above assertions (Stezano, 2018). At the moment, the models of AI in existence are based on the pattern of neural networks in the human brain. These neural networks take in many data and process them in line with the expected result. The machine then learns the relationship between the data and the desired effect so that in the future, it can easily apply the knowledge to new data to predict the result (Travaly & Muvunyi, 2020). But beyond this, there is deep reinforcement learning, whereby intelligent computers apply the method of trial and error to make decisions (Methods and goals in AI, 2022).

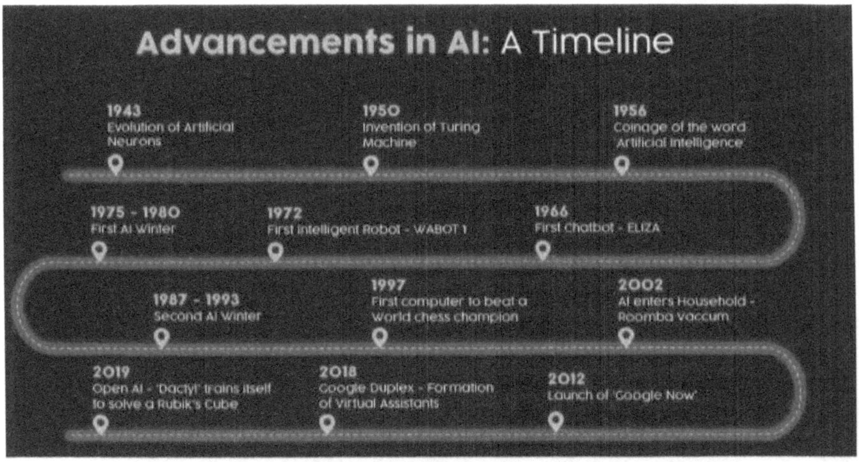

Figure 14: Advancements in AI from 1943 -2019
Source: Adapted from AI Timeline from the origin (https://pin.it/7IWNA0w)

Usefulness of AI

Artificial intelligence presents several benefits to modern life. It has become critical to virtually every area of life, offering various benefits to modern organisations. These benefits include:

Automation Leading to Fewer Errors: AI executes repetitive tasks automatically without experiencing fatigue which might lead to errors if humans were to handle the tasks.

Enhancement of Customer Experience: AI ensures a more fulfilling experience for end users of products and services. This is done via capabilities such as recommending products or services and optimising conversations.

Saves Time: AI reduces the amount of time taken to execute tasks. It makes it possible to perform many tasks simultaneously, reducing the time taken to get things done.

Analysis of Data: AI is employed to study large data sets faster and more accurately than humans. It can uncover and identify patterns that humans may likely miss.

Accurate Decision-Making: AI facilitates accurate decision-making by analysing data and accurately uncovering patterns and trends. This allows organisations to make better decisions resulting in improved business processes and more fruitful outcomes.

Areas of AI Application

Artificial intelligence has handy applications in numerous areas of life. These include:

AI Applications in E-commerce: AI is used in e-commerce to facilitate personalised shopping. AI-powered assistants and chatbots offer shoppers assistance that enriches their experience on e-commerce websites. Also, AI-enabled systems can detect and curtail credit card crimes and fake news.

AI Applications in Navigation: AI is useful for navigation. It enables users to successfully navigate movements by avoiding obstacles and figuring out how to move from one destination to another more efficiently and within the shortest possible time.

AI Applications in Robotics: AI-powered robots move and operate independently, executing various precision tasks such as carrying goods from one place to another.

AI Applications in Human Resources: Companies use intelligent software powered by machine learning to facilitate the hiring process. This software scans résumés and screens them reasonably and efficiently, increasing the chances of hiring the right people. AI also reduces the workload of recruiting teams by automating all mundane and repetitive tasks.

AI Applications in Healthcare: Artificial intelligence is being incorporated into medical appliances to enable them to detect diseases, manage chronic conditions, and facilitate drug discovery.

AI Applications in Agriculture: Using machine learning, computer vision, and robotics, AI can detect soil nutrient deficiencies so that farmers know which fertiliser to use. Also, AI can help agriculture workers analyse the soil, read the weather and understand the right time to plant. With precision farming and predictive analytics, AI gives farmers access to data that guides their farming activities to ensure optimum yield.

AI Applications in Gaming: AI creates smart, responsive, human-like non-player characters (NPCs) to interact with the players. AI enhances the game-player experience.

AI Applications in Automobiles: Car makers such as BMW, Tesla, and Volvo are incorporating AI capabilities to boost manufacturing, enhance supply chain efficiency, and make driving a safer and more comfortable experience for car users. Also, AI can be integrated into a car to facilitate easy navigation from one point to another.

AI Applications in Social Media: Facebook uses an AI called DeepText to understand better people's conversations on the platform. Many AI systems have become highly competent in understanding and translating text and even multimedia messages from one language to another. We notice a lot of this now on Facebook and similar social applications. Twitter uses an AI that recommends tweets that users might find helpful based on the kind of content they engage with on the platform. Also, Twitter uses AI to fight against hate speech in racist, sexist, homophobic or Islamophobic comments or tweets. Instagram uses DeepText to identify questionable posts and delete them to fight against hate speech and cyberbullying.

AI Applications in Marketing: With the help of AI, marketers can deliver ads that are highly personal and targeted at a particular audience based on an analysis of the pattern of behaviour of the audience. Also, AI helps with ad retargeting to ensure more optimal results.

2. Blockchain Technology

Blockchain technology is a system that keeps records of public transactions in various databases in a connected network via peer-to-peer nodes. The transactional record is also referred to as "block," while the databases are called the "chain" or "digital ledger." Transactions in the ledger can only be carried out when the owner authenticates it with a digital signature. Only with this authority will the transaction be allowed to take place, protecting the system from tampering. Hence, blockchain technology provides a highly secure system. Interestingly, the data can be seen by anyone, but no one can corrupt it.

How Blockchain Works

Three powerful technologies combine to make up blockchain technology. They are the Cryptography keys, a peer-to-peer network containing a distributed ledger, and a computer storing the network's records and transactions.

Cryptography keys comprise two types: the private crypto key and the public crypto key. These two keys aim to ensure that the transactions between parties are successful. An individual has access to two keys which they can use to generate a safe digital identity reference, otherwise known as a digital signature. With this, the individual can control and authorise transactions.

The digital signature connects with a peer-to-peer network so that when a deal is authorised, it automatically goes through a mathematical verification process before it is certified (Daley, 2021; Levy, 2021; Simplilearn, 2020). Through this process, a secured transaction is successfully implemented between the two parties that are connected to the network.

The Benefit of Blockchain Technology

The benefits of blockchain include the following:

Better Transparency: The shared ledger by blockchain allows it to record data and transactions simultaneously in multiple locations. The data and transactions are permanently recorded with a date and time stamp. Therefore, all the network participants can view the information in real-time. They can also view transaction history, reducing any chances of fraud.

Improved Security: By creating end-to-end encrypted records, blockchain provides a secure system against unauthorised or fraudulent activities. Privacy Issues can quickly be sorted on blockchain by safeguarding personal data and preventing unauthorised access through permissions.

Increase in Speed and Efficiency: Blockchain allows documents and transaction details to be stored in a system, eliminating the necessity for paper exchange. Also, the reconciliation of various ledgers is optional, thereby improving the speed of clearing and settlement activities.

Traceable Transactions: Blockchain allows for the sharing of data between individuals. An audit trail is created to document all transactions from their inception. This helps in tracing any data or transactions over time.

The Application of Blockchain

Given that blockchain operates as an immutable ledger of transactions that chains together blocks of data using cryptography, it bears many valuable applications, including in the following areas:

Financial transactions: Blockchain transactions are less expensive and faster. It allows participants have better control and security of their transactions. Although blockchain mainly deals with cryptocurrencies, its concept can be implemented in conventional investment.

Lending: Through the help of smart contracts, lenders can execute loans with collaterals on a blockchain. The smart contract system of the blockchain allows an automatic trigger of specific processes such as margin calls, service payments, collateral release, and loan repayment. The process is cheap and fast and offers lenders a reasonable rate.

Safe Sharing of Medical Information and Data: Record keeping on the blockchain allows medical practitioners access to up-to-date and current patient information. This can help ensure that patients who visit several doctors get the best healthcare as their medical data are easily pulled up.

Transfer of Money: Transferring money with the blockchain is faster and cheaper than most traditional transfers, especially in international transactions.

Ensuring Credible Voting and Security of Votes: Blockchain technology could ensure that no one votes twice. Therefore, votes are captured, transmitted and reported in a trusted manner, mainly as all of the transactions in a blockchain application can be carried out with anonymity and auditability.

Ensuring the Security of Personal Information: Blockchain can be used to secure access to personally identifiable information such as a National Identity Number and other such personal information so that only those who are authorised can access the information.

Insurance Transactions: With contracts and claims recorded on the blockchain, transparency will prevent customers from making double claims for the same event.

Providing Easier Verification Process for Real Estate Processing Platform: Blockchain technology can offer more security for real estate transactions, including financial transactions, ownership, and transfer of deeds and titles.

Artist Royalties/Music Royalties Tracking: Blockchain can prevent piracy of music and films since it does not allow a file to exist in more than one place. It can be used to track and trace files of music that are pirated and stored in a system that can be accessed through the Internet.

For Tracking Logistics and Monitoring Supply Chain: Logistics and supply chain partners can work more effectively together if they use blockchain technology for their operations. With blockchain, it will be easier for them to track items as they move through logistics. Blockchain will also ensure the integrity of data since data on the blockchain cannot be altered.

3. Internet of Things (IoT)

The Internet of Things (IoT) refers to a network of physical objects, such as humans, buildings, transportation systems, vehicles, farm animals, office appliances, and home appliances. Each corresponding object can be reached and communicated to as a result of sensors attached to them, which assists them in acting based on the instructions given to them, enabled by software (Matob, 2021).

A person carrying a heart monitor implant, a car with built-in sensors to detect low pressure in tires, and a farm animal with a biochip transponder are examples of physical objects or things that constitute IoT.

How an IoT System Works

Four distinct components comprise an IoT system: sensors, connectivity, data processing units, and user interface.

The sensors collect data in their environment based on prevailing circumstances. The data could be the level of substances, including liquids, powders, and granular materials present, the data on the overall temperature in the environment, or a full video feed.

The data is sent to the cloud via various connectivity, including cellular, satellite, Wi-Fi routers, Bluetooth or Ethernet connections, etc. Once the data reaches the cloud, it is processed to decide the prevailing situation in the environment from which the data was collected. Upon determining the prevailing circumstance – say the data processed indicates that the temperature was too high in the environment where the data was collected – the information extracted from the processed data is made applicable to the end user in one of two or more ways: The end user is alerted by a message sent to him through email, text message or a notification. Depending on the specific IoT application, the user may be able to remotely act on their phone to effect a change in the prevailing environmental condition (Edureka, 2020; Tech 21 Century, 2020).

On the other hand, the end user could proactively check on the environment via the IoT system, such as when a user reviews the video feeds in their house using their mobile phone, an app, or a web browser.

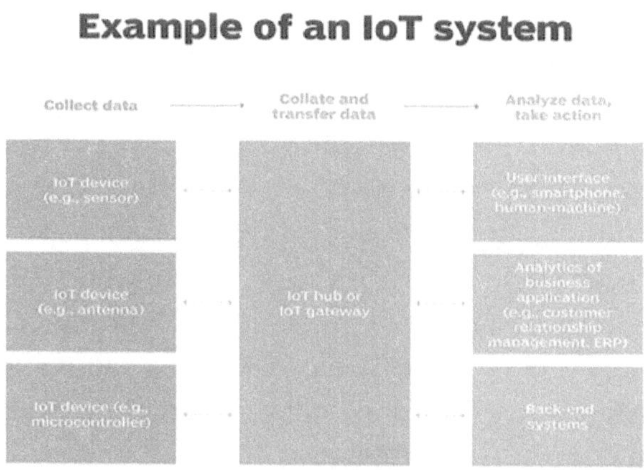

Figure 15: Example of an IoT system
Source: Online Design/TechTarget, TechTarget.com

Usefulness of IoT

As with many other exponential technologies, IoT enables people to control the gadgets and objects around them. It has helped individuals and businesses to gain control over their lives and activities in the following ways:

1. It offers individuals smart devices to automate their homes.
2. It allows businesses to look into their operations on a real-time basis; they can gain complete insight into such aspects as the performance of their machines, supply chain management, and logistics operations.
3. With IoT, businesses can automate some of their processes, thereby reducing the cost of labour.

4. IoT helps businesses to improve efficiency and thereby cut down on waste while increasing service delivery.
5. By cutting down costs, it makes businesses more competitive.

Areas of IoT Application

The Internet of Things (IoT) has applications in many areas of life, particularly in the following:

Wearables: They are devices such as GPS tracking belts, virtual glasses, fitness bands, and others which are fitted with sensors. When worn, they can be used to gather data about the user's physiological functioning, which helps to detect any abnormality early and seek to correct it immediately.

Application of IoT in Health: In many hospitals, devices with embedded sensors are attached to patients, allowing doctors to remotely monitor high-risk patients' conditions. Doctors can give patients optimal and personalised care by continuously monitoring specific metrics via the data-embedded sensors. Also, in some hospitals, beds are embedded with sensors enabling doctors to observe and assess blood pressure, body temperature, and the performance of vital organs such as kidneys, the heart, and lungs. In addition, many hospitals now use IoT to complete inventory management for medication and medical instruments.

Application of IoT in Fleet management: With sensors installed in fleet vehicles, managers can keep tabs on every detail of the cars, including their status and maintenance needs, by simply accessing software that collects and processes data about them. Also, the application of IoT can help with geolocation, performance analysis, and fuel savings via telemetry monitoring and control.

Application of IoT in Agriculture: With the incorporation of IoT, farmers can gain detailed insight into the soil conditions where they grow their crops, such as soil moisture level, acidity, nutrients present or absent in the soil, soil temperature, amount of light, humidity, and others. This knowledge is crucial in helping farmers determine which crops to grow, whether irrigation is required, and which fertilisers to apply.

Application of IoT in Hospitality: IoT has a universal application in hotels. With IoT applications, keys to hotel rooms can be sent directly to guests' mobile phones. Also, through integrated IoT, guests can order room service and the supply of personal hygiene essentials. And when guests have checked out, doors to rooms are automatically locked while information is immediately offered regarding the available spaces.

Application of IoT in Smart Grid and Energy Saving: By using meters embedded in them and incorporating sensors at various points between the power generating plants and the distribution points, data can be obtained about the electrical network allowing for the detection of faults and repairs. The data gathered can also offer end users valuable information regarding their consumption patterns to manage their electricity expenses better.

Application of IoT in Water Supply: By incorporating smart meters in the water supply network, data is gathered, which supplies information on the entire network so that faults in the supply service are detected early, and a course of correction is initiated immediately. Also, consumers can monitor their consumption patterns in real time, making it easy to see when their consumption is out of range.

Application of IoT in Smart Buildings: The application of IoT in buildings involves installing sensors in buildings to help manage energy costs. For instance, sensors can detect the temperature of a room and automatically turn on or turn off the air conditioner; or automatically turn off the heat after sensing the absence of anyone in the house.

Application of IoT in Smart Cities: IoT can be applied in cities with large populations to help manage traffic, transportation, energy distribution, water supply, urban security, and others. By incorporating sensors in all these systems, city managers can easily monitor each system to ensure smooth and uninterrupted flow.

Application of IoT for Home Security: IoT is increasingly being employed in providing smart security for homes. The system works by connecting various IoT-enabled products, including doorbells, alarms, cameras, lights, locks, and microphones, with each other to offer security for the home. For instance, the system can manage surveillance around the house, and if someone rings the doorbell, the homeowner can see and speak to the visitor via their mobile phone. The homeowner can remotely monitor and manage the system via their mobile phone from any location.

Application of IoT for Industrial Security and Safety: Surveillance cameras embedded with sensors can monitor the perimeter of industrial complexes and restricted areas within such premises. Sensors can also detect the presence of hazardous chemicals or substances resulting from leakages or pressure build-up. Such leakages can quickly be addressed before they result in damaging outcomes.

The Future of IoT

From 2008 to 2009, when the number of connected devices first outnumbered the global human population, the growth of IoT devices has steadily risen. According to a report by Juniper Research (2021), there were about 46 billion connected devices around the world in 2021. This is an increase of nearly 200%, which is relatively high, as was the case in 2016. Security Today (2020) estimated that by the end of 2021, 31 billion IoT devices would be installed worldwide. It is now speculated that interconnected IoT devices will increase to about 125 billion by 2030 (Martech Advisor, 2019).

Statista (2022) estimates that there were only 10 billion IoT devices in use in 2020, based on connections in the United States alone. Still, by 2030, every IoT device will be interconnected to at least ten other IoT devices, with the market for IoT reaching a warping $500 billion by 2021.

While there are several players in the IoT devices market, Futurum Research (2022) has predicted that by the end of 2022, Google Home will have the largest market share at 48%, followed by Amazon Alexa at 37%, and, finally, Apple HomePod at 12%.

In terms of connectivity, IoT World Today (2021) stated that attention would shift from applying 5G technology to IoT. Instead, low-power networking systems and satellites would grow by 20% throughout 2021. This would give homeowners and companies more options to choose from. Some companies – for instance, lighting and factory machinery– could choose different connectivity options based on what they consider more cost-efficient, time-saving, and safe.

Statista (2022) predicts that by 2022 the smart home market will increase to $54 billion. Users will use the technology to control the gadgets and switches for home appliances and entertainment. Intel (n.d.) predicts that by 2025 the worldwide market for IoT devices will rise to $6.2 trillion. It is also expected that the quantum of data generated through IoT will reach 79.4 zettabytes, and using the same data will attract 80 zettabytes of data.

By 2023, networked device ownership worldwide will reach 3.6 per person (Cisco, 2018). Cisco also predicted the global ownership of IoT devices to be as follows:

- 13.4 per person in North America
- 3.1 per person in South America
- 9.4 per person in Western Europe
- 4.0 per person in Eastern and Central Europe; and
- 1.5 per person in Africa and the Middle East.

Summarily, in a few years, almost all types of infrastructure will become interconnected by IoT, including civil engineering, construction, road, transport, industrial and civil aviation infrastructure. It will not only constitute devices used at home; it will be made up of both devices used at home and machines used in factories, shops, hospitals, schools, and other such places.

4. Robots

Robots are programmable machines built to mimic human or animal actions or support humans in performing specific activities. The science involved in making robots is called robotics, an intersection of science, engineering, and technology.

Originally, robots were built to handle monotonous or repetitive tasks such as making cars on an assembly line (Harris & Pollette, 2002). Since then, more advanced robots have been built to handle such tasks as fighting fires, cleaning homes, assisting with surgeries and exploring some of the harshest places on earth.

Main Components of a Robot

Robots are built to perform various tasks akin to what humans would do, requiring various specialised components to complete these tasks. Nevertheless, specific components are central to the construction of any robot, no matter what task it is designed to fulfil. Generally, robots comprise five main components, namely:

1. Control System
2. Sensors
3. Actuators
4. Power Supply
5. End Effectors

Control System: This can be considered the brain of the robot. It is programmed to tell the robot how to utilise its various parts in the same manner the human brain communicates to different body parts through signals to carry out specific tasks. All the parts of the robot that make up its central processing unit constitute its control system. All robots contain a measure of computer programming within the central processing unit that tells them what action to execute or which tasks to carry out and when.

Sensors: These parts enable a robot to detect stimuli in its environment. Robots detect these stimuli as electrical signals processed by the controller. Some sensors found within robots are video cameras, which function as the eyes. Other sensors are photo-resistors or photovoltaic cells, which respond to light; proximity sensors, which detect the presence of nearby objects without any physical contact; temperature sensors, which detect the change in the temperature of a surrounding; acceleration sensors, which are used for measuring acceleration and tilt; and sound sensors such as microphones that serve as ears. The combination of these sensors equips the robot with the capacity to detect stimuli in its surroundings and determine the most appropriate response after the control system has processed the stimuli and relayed the right course of action. *For a simple obstacle avoidance robot, photo-resistors can serve as sensors. However, more sensors would be required if a more complex robot is built.*

Actuators: Actuators are devices within a robot that enable various parts to move. They activate the joints of a robot's arm to rotate, the robot's wheels to turn, or the robot's gripper to open. The movement facilitated by actuators occurs due to motors embedded in the devices. The engines execute instructed actions due to signals from the control system. Actuators could be made of elastic or metal materials, either pneumatic actuators (operated by compressed air) or hydraulic actuators (operated by oil).

Power Supply: Robots require a source of power to function. For robots that move about, the most typical energy source for their operation is internal batteries, mostly lead-acid batteries or sometimes silver-cadmium batteries. Like those used in factories, stationary robots are usually built to run on AC power. It is believed that in the future, robots will be made to run on pneumatic power from compressed gases, solar power, hydraulic power, flywheel energy, and nuclear energy.

End Effectors: These components that robots are equipped with allow them to carry out their tasks. Robots in factories are usually equipped with features such as paint sprayers and drills. In contrast, robots in the operating theatre are equipped with components such as scalpels and other surgical tools. Robots built for packing and delivering goods or, say, for the diffusion of bombs, come with gripping claws.

Uses of Robots

Over the years, the use of robots has gradually expanded to include virtually every field of human endeavour. And given the technological changes, it is likely that, shortly, we will have robots working side-by-side with humans, executing tasks everywhere, including hospitals, hotels, schools, and event centres. And even on our roads, directing and controlling traffic (Muthoni, 2021).

Some of the uses of robots include the following:

Robots Are Used in Manufacturing: Robots have been assisting humans in manufacturing for quite some time now. They are used on the assembly line to assemble products such as cars and industrial equipment and to produce various consumer goods.

Robots Are Used for Logistics: Robots are now employed in handling and shipping packages. On the floors of warehouses, robots pick up items from shelves, move them across the warehouse floor and package them. In the near future, there will be robots that will autonomously deliver packages to people's doorsteps. In hotels, guests will soon have their meals delivered to them in their rooms by last-mile robots.

Robots Assist with Home Chores: In many homes, robots are seen helping with house chores – such as vacuum cleaning, mowing the lawn, and cleaning the pool. Also, robots offer reminders of schedules and even provide entertainment.

Self-Driving Cars: Giant automakers such as Tesla, BMW, Ford, and Volkswagen are leveraging data science and robotics to build cars that would not require human intervention to navigate from one point to another. With these self-driving vehicles, humans will sit back, relax, and enjoy the ride while the car moves by itself.

Healthcare: In most developed parts of the world today, robots are used in healthcare management and operations. From robots that assist in surgeries to those that move through hospital wards delivering medications and bots that help with physiotherapy, robots have been extensively used to provide top-notch healthcare services. Since the outbreak of COVID-19, robots have been used in medical and pharmaceutical laboratories to seal COVID-19 testing swabs. Some medical equipment manufacturers have employed robots to produce PPE and respirators.

Robots Are Used for Bomb Diffusion and Detecting Landmines: Generally, robots are used for tasks that are considered dangerous to humans, such as going into a building that might have a bomb in it or an area that is suspected of having landmines buried in it. Also, during search and rescue operations, robots are deployed to handle jobs in regions that are too risky for humans to venture into and to undertake tasks that are too hazardous for humans to carry out.

Robots Are Used for Exploring: Robots are used for exploring underwater, other planets, and the harshest parts of the earth.

5. Big Data

As earlier pointed out, data refers to discrete symbols, characters or symbols that a computer can manipulate. Traditionally, they have been successfully handled by data processing systems, which sort, process, and store them using optical, magnetic, and other recording devices. In today's world, many systems, organisations and countries generate enormous amounts of data that are so huge that they are impossible to manipulate with traditional data processing systems. Such a vast quantity of data is called "Big Data." So, big data refers to vast amounts of data of enormous quantity and volume, growing rapidly (UpGrad Blog, 2020; Taylor, 2020).

From this definition, some key characteristics of big data are apparent. These are volume, velocity, and variety, and as pointed out above, they come with high speed, in huge volumes, and often from varying sources.

Volume: As the name suggests, volume is a crucial attribute of big data. The volume of data generated is essential in determining if the particular data can be considered big data or otherwise.

Velocity: About big data, momentum refers to the speed at which data is churned out or generated. It depends on the rate data flows from various sources – such as sensors, application logs, social media platforms, mobile devices, and others. The extent to which data is speedily generated and processed to make inference determines the actual value of the data. RFID tags, smart metres, and numerous IoT devices stream data into businesses rapidly and must be processed promptly.

Variety: Data comes in structured and unstructured formats and from various sources – spreadsheets, databases, numbers, emails, photos or images, video feeds, surveillance devices, audio recordings, etc.

Examples of big data include data from the following sources: social media networks, customer databases, emails, medical records, mobile apps, financial transaction processing systems, data from sensors on machines, and industrial equipment and IoT devices.

Types of Big Data

There are three types of big data. They come in the form of structured data, unstructured data, and those that can be classified as somewhat structured and somewhat unstructured, i.e., those that are semi-structured data.

Structured Big Data: When data is well organised in a manner that is easy to be stored, accessed and manipulated, it is structured big data. It is usually stored in databases and is the type of data most people deal with regularly. Examples of structured data include numbers, dates, groups of words, and names called "strings." Experts claim that structured data comprises about 20% of the data.

Unstructured Big Data: Here, the vast data is disorganised and difficult to be manipulated straightforwardly. Usually, unstructured data is heterogeneous, containing a mixture of text files, images, videos, dates, numbers, and facts. The heterogeneous nature of unstructured data results in irregularities that make it difficult to process with traditional data processing programmes. This type of big data is roughly 80% of all the data produced by institutions worldwide today.

Examples of unstructured data include satellite surveillance imagery, high energy physics, atmospheric data, seismic imagery, traffic video feeds, oceanographic seismic profiles, and data from social media platforms such as YouTube, Facebook, Twitter, LinkedIn, text messages, location information, and others.

Semi-structured Big Data: This type of data is neither wholly structured nor completely unstructured. It is thus in-between the two types of data. It is the type of data that has some consistency to it. However, it also contains some inconsistency and variability and does not conform to a rigid structure like the kind that is needed for relational databases. Photographic images and delimited files, as found in, say Microsoft Excel records are examples of semi-structured big data.

Application of Big Data in Industry and Government

Big data has a variety of applications. They include the following:

Application of Data in Customer Care: Big data gives businesses much insight into their customers' shopping behaviour to resolve all potential difficulties at every touch-point, even before customers bring them up. Businesses can analyse data to determine customers' preferences, enabling them to be responsive in real time.

Application of Data in Marketing: Big data provides businesses with valuable insights that allow them to target their marketing and advertising efforts at those segments most likely to purchase their products. This way, they increase engagement as well as conversion rates.

Application of Data in Improving Business Operations: Data gathered from various sources in a business can be utilised to improve its physical operations. For instance, machine data can be used to schedule predictive maintenance to prevent costly repairs and downtime.

Application of Big Data in the Healthcare Sector: Doctors and medical researchers now use big data to identify the signs of certain diseases and risk factors and diagnose ailments in patients. With the abundance of patient data, medical scientists have more records to research diseases such as cancer and Alzheimer's. Also, big data allows the government to determine the allocation of resources designated to fund healthcare delivery without incurring wastage.

Application of Big Data in Finance: Big data and predictive analytics enable big banks, hedge funds, insurance companies, and other players in the finance industry to curb fraud, undertake a risk assessment, carry out trade analytics, conduct credit rankings, and offer brokerage services. Financial organisations are some of the most significant users of big data today.

Application of Big Data in Agriculture: Big data is employed in seed engineering and crop yield prediction. Many farmers, organisations and governments now use this technology to determine and keep an eye on the different types of seeds, crops, and livestock they have at every point in time, even in a whole country, because this technology can deal with vast amounts of data.

Application of Big Data in Transport: By leveraging big data, managers at government transport agencies can analyse traffic flow on different roads in cities and urban centres and identify any potential threat, to smoothen traffic flow. The potential threat could be averted by re-directing traffic away from the areas in real time.

Application of Big Data in Education: Analysis of big data can assist organisations and governments in determining the educational interests, needs and deficits of their society. This makes it easier for government to formulate policies that will ensure that all young people receive the highest quality of education.

Application of Big Data in Weather Forecasting: Weather agencies use big data to forecast weather conditions. They collect a large volume of data they analyse to create the right weather conditions for regions and the entire country.

6. Cybersecurity

Cybersecurity refers to the systems and techniques involved in protecting computers, networks, software, programmes, data, and servers from unauthorised access or malicious attacks for exploitation or to cause havoc (Rouse, 2011; Kaspersky, 2022).

Elements of Cybersecurity

Cybersecurity solutions should encompass the following elements:

- Application security
- Cloud security
- Data security
- Endpoint security
- Identity management
- Critical infrastructure security
- Database security
- Mobile security
- Network security
- Operational security, and
- Disaster recovery/business continuity planning

Types of Cybersecurity Threats

There are multiple cybersecurity threats. The following are some of the more common ones:

DDoS Attack: Distributed Denial of Service Attack occurs when cyber criminals send too much traffic to a network or its servers and, thus, overwhelm it. As a result, the network is rendered unusable, so it cannot process legitimate requests from within the organisation that owns the network.

Malware: This is the most prevalent type of cyber threat that individuals, organisations and governments suffer. Cybercriminals create malicious software to disrupt or damage other people's computers. They are often spread through unsolicited email attachments or downloads that look legitimate and harmless.

The different types of malware include:

a. Virus – Viruses can replicate themselves. They enter a person's system, embed themselves onto a workable file, and after some time, start to repeat themselves in such a manner that they corrupt the host file, the system and other interlinked files or systems.

b. Trojan – This is malware disguised as genuine software that, when uploaded, causes actual damage to the system or allows the criminal to extract data from a network.

c. Spyware– This malware is designed to record what a user does or any personal details they enter on their computer, such as credit card details.

d. Ransomware – Ransomware is usually sent by Internet criminals to a person's, organisation's or a country's system or database to overwhelm the system so that they get bogged down and unable to function at all or optimally. Until such malware is deleted, the data and files on the computer are at risk of erasure or losing integrity.

e. Adware – This software displays pop-up adverts on your computer or mobile device. Usually, adware is downloaded along with a free computer programme or app.

f. Botnets – This refers to a network of computers that cybercriminals have infected with malware and taken over their control. Such computers are remotely manipulated and controlled by cybercriminals to execute any action they want, such as financial gain, launching an attack on another network, or harvesting individuals' details such as passwords, National Identity Number, credit card numbers, addresses, telephone numbers, etc.

g. Man-in-the-Middle Attack – This cyberattack occurs when a criminal intercepts communication or data transmission between two or more people. This happens mostly when the individual communicating or exchanging data uses an unsecured Wi-Fi network.

h. Phishing – This is when criminals send victims fake emails that appear to be coming from a legitimate source to obtain sensitive or personal information such as credit card details or National Identity Number from the victim.

i. Social Engineering – This is the type of cyberattack in which criminal attempt to use impersonation to trick their victim into giving up sensitive information or personal data or to extract money from them.

j. SQL Injection – In a Structured Query Language (SQL) cyberattack, a criminal exploits vulnerabilities in a data-driven application to introduce malicious code in a database via a malicious SQL statement. The presence of the malicious code allows them access to the sensitive information in the database.

Benefits of Cybersecurity

There are numerous benefits of implementing standard cybersecurity practices in government and business offices. Some of these benefits include the following:

i. It protects networks and data from unauthorised access.

ii. It improves information security and business continuity management.

iii. It inspires stakeholders' confidence that their data will not be compromised.

iv. Improves company credentials with the proper security controls in place.

v. It enables faster recovery time in the event of a breach.

vi. It keeps organisations' websites up so they do not crash.

vii. It increases the productivity of computers and maximises their output.

7. 3D Printing

3D printing is an additive process of manufacturing in which a computer is used to create a 3-dimensional object by placing one layer of material upon another until the entire object is created. 3D printers print by placing thin slices of their materials above one another until the 3-dimensional object they are printing takes shape. 3D printing is a departure from the traditional subtractive manufacturing process in which material is removed from a larger block of fabric to create a final design (Hayes, 2014a). Plastics, aluminium and steel, other metals, resins, ceramics and even woods are some materials used in 3D printing.

Different Areas 3D Printing is Used in Industries

These are some of the many areas 3D printing is applied in industries today:

Aerospace Industry: In Aerospace, many sub-components and parts of a plane, for example, are produced with the help of 3D printing without reliance on huge factories. This significantly reduces wastage as well as lead time.

Automotive Industry: The automotive industry employs 3D printing to create prototypes of parts to be manufactured or to produce parts that are no longer available, like needed spare parts.

Medical & Biomedical Engineering: Implants and other medical devices are today being produced with the help of 3D printing. A scan of the body part to be fitted with a medical device is taken, after which 3D printing is employed to create the device to work accurately with the scanned image. Customised prosthetics, like limbs and organs tailored to match a patient's body, are produced using 3D technology.

Rail Industry: 3D printing has been used in the rail industry to repair worn-out rail tracks. In addition, it is used in creating customised armrests for train drivers and housing covers for train couplings.

Robotics: In the robotics industry, 3D printing is used to create bespoke exoskeletons and responsive robots with an increased ability to move swiftly and more efficiently.

Building and Construction Industry: Operators in the building and construction industry use 3D printing to make models of the layout or shape of buildings to be constructed or created. Architects and building engineers are exploring and leveraging 3D printing to create entire buildings.

Food Industry: Recently, 3D printing has emerged as a modern method of preparing and presenting food. In the food industry, 3D printing was first applied to foods such as sugar and chocolate. Lately, it has progressed to the printing of meat at the level of cellular protein. Most of our foods and food ingredients, like different types of meats, will be 3D printed.

Logistics and Supply Chain: Many customer-required products can be 3D-produced, with many customisations, using 3D technology. This eliminates the need to stack the shelves of grocery stores with finished products. This reality completely collapses the supply chain to its basic form.

Mass Customisation: With 3D printing, products can be manufactured at once to fit the needs and requirements of various individuals without incurring any additional cost.

Other applications of 3D printing are in the design of fashion, sports shoes, jewellery, arts, and sculptures.

8. 4D Printing

A new type of additive printing is the latest 4D printing. Even though 3D printing is still popular, new materials, applications, and procedures for 3D printing are being developed.

4D printing is the process by which an object printed with 3D technology transforms itself into another structure due to the influence or impact of water, light or heat over time. This means that 4D printing is a 3D-printed object changing shape over time. The materials used in 4D printing are programmable and advanced materials that change shape when exposed to water, light or heat.

How 4D Printing Works

The diagram below illustrates how 3D and 4D printing work. 3D-printed objects have a fixed shape. This means that once printed, the condition of such objects will never change later.

The shape of 4D printed materials can change. This is why they are also called "smart materials," using materials like hydrogel and polymers that have a memory of their previous shapes. As a result of the thermomechanical properties of these materials, objects printed with them can change shape over time when exposed to water, light and heat.

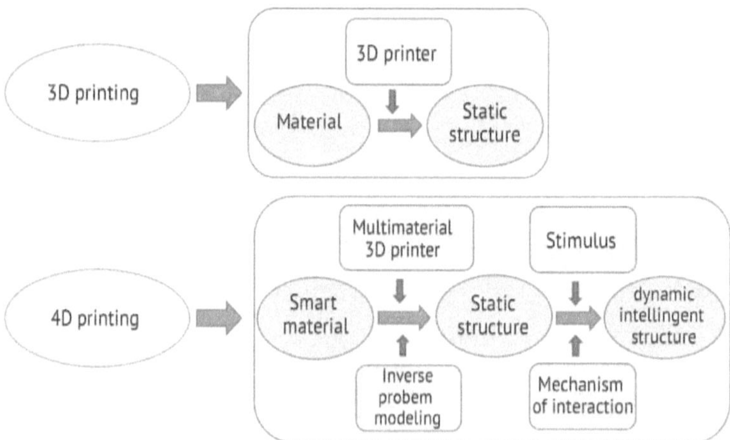

Figure 16: *Picture of How 4D Printing Works*
Diagram adapted from [F. Momeni et al. / Materials and Design 122 (2017) 42–79], re-drawn by Jean-Claude André | Scientific Advisor at INSIS

Potential Areas of the Application of 4D Printing

4D printing has so much more to offer the world about how we will produce many future products. These exciting potentials include the following:

Aerospace Industry: 4D printing can make structures that deploy themselves in ventilation systems and engines. Researchers believe that the 3D printed tensegrity structures – called "struts" – are made from shape-memory polymers, which quickly fold flat but can unfold and expand when exposed to air and can be used as an antenna in spacecraft.

Defence: Research is being conducted to create a military uniform that can alter its camouflage in contact with poisonous gases or shrapnel, thus offering better protection to soldiers. Also, researchers are exploring the possibility of building self-assembly bridges and shelters for military use.

Medical and Health Industry: 4D printing is being explored in tissue engineering. The concept involves using biocompatible hydrogels to print artificial skin that would be grafted into parts of the body that have lost skin. Such implants could also be developed with 4D printing. Another potential application of 4D printing in medical science involves designing programmable stents that would travel through the human body and open up when they reach their destination.

Automotive Industry: Engineers at BMW and MIT Self-Assembly Lab are working on possibly employing 4D printing in the future car. BMW is working on building a 4D printed car, "VISION NEXT 100," which will have an outer shell that could change in response to air flows.

Consumer Goods: 4D printing could make consumer goods like cups, cutleries, pots, and pans.

Self-Repair Piping System: It is anticipated that pipes used in plumbing systems will soon be built with 4D printing to change diameter in response to the flow rate and water demand. Also, the pipes could heal themselves automatically whenever there is a breakage or a crack, given their ability to alter their shape in response to environmental changes.

Fashion Industry: 4D printing could be used to make clothing that could alter based on the weather condition or the activity of the individual wearing the clothes. Shoes and other footwear could also be made, expanding when the wearer is running. The expansion will afford more room to ensure comfort.

9. Additive Manufacturing (AM)

Additive manufacturing is the set of technologies that use data, computer-aided-design (CAD) and software, or 3D object scanners to manufacture objects sequentially by depositing material layer upon layer in precise geometric shapes (Spatial Team, 2020; GE Additive, n.d.; Zero Hour Parts, 2020).

This manufacturing method is referred to as "additive" because the material is added sequentially to create an object.

Additive manufacturing is the opposite of traditional manufacturing, whereby material is continuously removed from a solid block until the shape of the desired object is created, or separate components are joined until the desired object is created.

In manufacturing, we carry out these process steps:

- Design of product and process
- Carry out vibrational analysis
- Do iteration where needed
- Go through the manufacturing process
- Test the product, and
- Iterate again if necessary

With additive manufacturing, these manufacturing processes are used:

- Design of the product and process
- Carry out agile processing to determine the quality of the design and printability
- Do iteration where needed, and
- Go through the manufacturing process

Considering the time taken to manufacture parts using additive manufacturing, it is likely that it will only partially replace traditional manufacturing. Understandably, additive manufacturing is becoming the preferred method of manufacturing small batches of more complex parts, but conventional manufacturing will remain relevant for manufacturing simple and easy-to-produce items.

Types of Additive Manufacturing Processes

In 2010, the International Organisation for Standardisation created categories and explained the different additive manufacturing processes. Depending on the component or product to be used or created, any of the following additive manufacturing processes could be employed:

VAT Photopolymerisation: VAT Photopolymerisation, also known as stereolithography, results from using a vat of liquid photopolymer resin. A platform is lowered from the top of the resin, and the laser beam is used to draw the shape, creating a product layer to be manufactured. After this, Ultraviolet (UV) light is used to cure or harden the layer.

Directed Energy Deposition: The nozzle of a four- or five-axis arm deposits material on the existing surfaces of a stationary object. After deposition, the material is melted using either a plasma arc, an electron beam, or a laser. Subsequent layers of material are deposited and solidified, adding or fixing new material characteristics to the finished material. The material is available in wire and powder forms.

Binder Jetting: With the help of a roller, a powder-based material is used on a build platform, and after this, the print head is used to apply the binder on top of the forum. The binder, usually in a liquid form, glues the layers together. After each layer is created, the object is lowered onto the platform, and the next layer is made.

Material Extrusion: In this additive manufacturing method, the material, usually thermoplastic, is heated and then drawn through a nozzle and deposited to form layers to create the product. Since the material is melted, each layer deposited fuses effortlessly with the previous layer.

Material Jetting: This additive manufacturing method deposits the material as droplets solidifying to form layers. The parts then go through a process of hardening.

Sheet Lamination: This process is relatively low-cost. The material is placed on a cutting bed where layers are deposited in a manner that makes them bond to the material. The desired shape is cut out using a knife or laser.

Powder Bed Fusion: This involves mantling and fusing powder layers on the platform, using either a laser or electron beam. Layers are created successively to form the product.

Five Industries Where Additive Manufacturing Has Significant Application

Additive manufacturing has applications in many industries, including the following:

Aerospace Industry: Through additive manufacturing, aerospace engineers now design and print aerospace components that are complex and possess much strength but are also light and more efficient. This lowers the fuel consumed by aircraft and reduces the quantity of carbon dioxide emitted.

Automotive Industry: Using additive manufacturing, automakers can make complex, high-quality car parts that perform well even under harsh and highly demanding conditions. Also, with AM, automakers can prototype rapidly to build parts that will make cars safer and more comfortable.

Medical Manufacturing Industry: Some of the applications of AM in the medical field include the following: creating the anatomy of patients for surgeons to use for practice before the actual surgery, making prosthetics for patients, manufacturing customised implants, and making dentures and dental implants.

Energy Industry: Additive manufacturing allows the energy sector to create lightweight, efficient, and environmentally-friendly components that can perform well under severe conditions. Also, with AM, corrosion-resistant components can be developed for those parts that may be underwater.

Consumer Products Industry: Product design teams can leverage AM to rapidly prototype, iterate, pivot and quickly develop products that suit the market's demand or customers' needs.

10. 5G

5G is the 5th generation mobile network designed to meet modern society's need for connectivity and handling large amounts of data. With 5G, we have networks that deliver excellent reliability, enhanced availability, more capacity, peak voice and data speeds, low latency, faster connectivity and a better and more consistent user experience. Thus, in terms of latency, the amount of time it takes one device to respond to other devices is significantly reduced, from 100 milliseconds for 3G, to 30 milliseconds for 4G to 1 millisecond for 5G (EMF. n.d.; Gillis & Gerwig, 2022; Qualcomm, 2017). This is awesome.

Evolution of Generations of Mobile Networks

The mobile network we enjoy today has evolved through generations of invention and reinvention. So far, there have been five generations, namely, 1G, 2G, 3G, 4G, and 5G. The latest, of course, is 5G. 6G and 7G are also being developed, but the latest and most developed today is 5G.

The first-generation 1G was invented in 1980 to deliver analogue voice. The second generation, 2G, was developed in the early 1990s to introduce digital voice, such as CDMA (Code Division Multiple Access). The third generation 3G, which came into usage in the early 2000s, brought mobile data, while the fourth generation 4G LTE, which evolved in the 2010s, ushered in the era of mobile broadband. Computer scientists designed 5G to provide more connectivity than has ever existed (Qualcomm, 2017).

How 5G Will Shape the Future

As an integrated, more efficient air interface, the 5G mobile network can enable the user experience and deployment models not implemented on previous generations of mobile networks. With higher connectivity, reliability, and practically negligible latency, 5G can help next-generation possibilities within the mobile ecosystem (Remmert, 2019). Some of these possibilities include the following:

Growth of IoT: 5G will enhance the connection of billions of devices, including medical implants, refrigerators, lights, electronics, cars, advertising hoardings, transponders, etc., that will exchange data without human intervention. With low latency and high-speed connectivity, 5G will allow new possibilities for applications and services that have never existed before. Almost every device could become "smart".

Low Latency and High-speed Connectivity: With a low latency of 1 millisecond, 5G will provide far greater data speed and connectivity. It will make it easier to maintain real-time control or monitor devices, industrial machines, robots, vehicle-to-vehicle communication and safety systems, medical procedures, and surgeries.

Enhancement of Mobile Broadband: With 5G, there will be new applications for fixed wireless Internet access for homes and offices. There will also be outdoor broadcast applications without the need for broadcast vans, while people on the move will experience the ease of connectivity and incredible speed.

Smart Communities: With the ability to connect billions of devices, 5G will make it possible to develop smart communities at various levels, including smart cities, smart schools, smart homes, and smart cars.

Smart Factories: Another practical application of 5G is the creation of smart factories, which will bring about more efficient manufacturing processes through automation and save costs for manufacturers.

Automation will eliminate the possibility of human error and improve the quality of products. Smart factories will also create flexibility around customised products to meet individual customers' needs without incurring additional expenses.

Data-Driven Decision-Making: 5G and IoT will provide businesses and government agencies with much data that has never been available. Such a wealth of data will afford deep insight into operations and other aspects of these businesses and agencies. As such, critical decisions in these organisations will be driven by data, with the likelihood of more positive outcomes.

Autonomous and Connected Vehicles: 5G will compellingly revolutionise transportation. There will be the emergence of connected cars and autonomous vehicles, both of which will be able to operate effectively with a tremendous reduction in latency and a significant increase in network speed. Vehicles will be able to communicate or share data, thereby avoiding potential collisions and other such road accidents. Autonomous or self-driven cars can navigate seamlessly from one point to another with 5G, IoT, and AI.

Remote Surgeries: With 5G, it will become possible for doctors to perform surgeries remotely from anywhere globally by controlling and directing medical robots to perform procedures.

Accessibility of Virtual Reality (VR) and Augmented Reality (AR): With 5G, VR and AR will become ubiquitous and easily accessible. With 5G-powered VR, an individual could travel to a city and watch their favourite opera singer perform live with the feeling of being on the grounds. People can inspect a piece of real estate they are looking to buy and even walk through the inch of the building while sitting in their offices or relaxing at home.

11. Quantum Computing

Quantum computing is an area focused on developing computer technology based on the principles of quantum theory. With quantum theory, we can understand how critical physical properties of nature (matter and energy) function at the atomic and sub-atomic levels. A quantum computer uses quantum mechanics to enable it to perform certain kinds of computations that a regular or classical computer cannot perform.

Traditionally, regarding classical computing, information is encoded in bits that can either be 1 or 0 at every point. Thus, they can undertake operations at the core hardware level using the 1s and 0s. On the other hand, quantum computers use quantum bits or qubits that might represent 1 or 0, a combination of the two, or even a number between 1 and 0 simultaneously. For example, when you have two qubits in superposition, they mean that four possible outcomes can be determined, thus [0, 0]; [0, 1]; [1, 0]; or [1, 1].

A situation where four possible outcomes can exist simultaneously is referred to as supposition. The ability of quantum computers to combine qubits withholding all possible configurations of information at once, means that users can represent complex problems far easier than attempting to do the same on traditional or classical computers. It also means that a quantum computer will take significantly less time to crunch a data set (CB Insights Research, 2021; Frankenfield, 2018; FutureLearn, 2021).

In addition to supposition, entanglement is another crucial part of the quantum computing phenomenon. Just like in physics, quantum entanglement in computing occurs when some qubits are entangled, such that whatever happens to one happens simultaneously.

The above attributes of quantum computers are always more powerful than traditional computers of similar sizes; see **Figure** 17 for a detailed schematic. They can simultaneously carry out similar, and even reversible sets of computations and reach similar results.

Whereas a traditional computer can carry out one operation at a time, a quantum computer can simultaneously carry out similar and even different operations in several computers as it uses the same processing unit.

While classical computers are better at tasks, including email, spreadsheets, and desktop publishing, quantum computers are well-suited for cryptography, modelling, and indexing vast databases (Vogel, 2019; CB Insights Research, 2021).

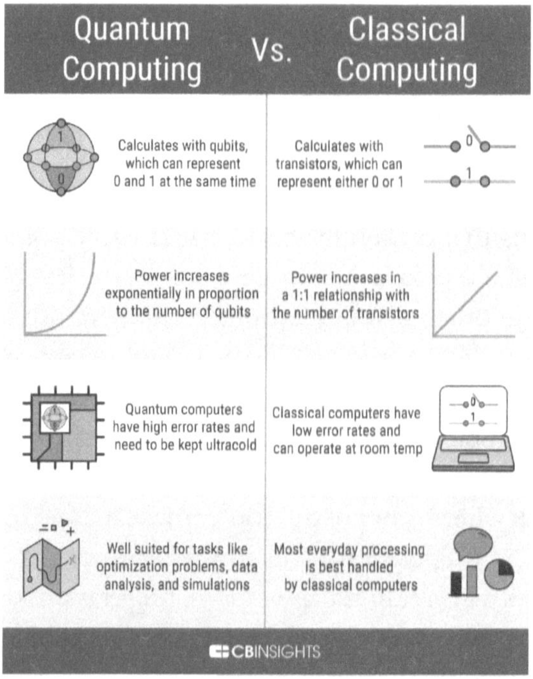

Figure 17: Quantum vs Classical Computing
Source: CB Insights Research

Advantages of Quantum Computing

The unique computational power of the quantum computer confers on it certain advantages. They include the following:

They Are Fast: After conducting a speed comparison between classical and quantum computers, Catherine McGeoch (n.d.) found out that quantum computing is a lot faster by solving the same problem on each computer. In 2019, Google claimed that its 54-qubit sycamore processor calculated random numbers in about

200 seconds, which would take the most powerful supercomputer around 10,000 years (Cossins, 2019). Such analysis is impossible on a traditional or classical, non-quantum computer.

Solving Complex Problems: Due to the ability to handle supposition and entanglement, quantum computers can run computations with a much higher order complexity, no matter the level of combinatorial explosions involved with the operating variables. Each day generates about 2.5 exabytes of data. This figure is equivalent to the content of 5 million laptops. Only quantum computers with their computational capacity can process such an amount of data. Therefore, as quantum computers have enormous computing power at high-order magnitudes, they will always surpass classical computers in processing power.

They Can Run Complex Simulations: Due to the speed with which they operate and their level of complexity, quantum computers can simulate many intricate systems.

The Uses of Quantum Computing

As quantum computers are good at solving certain kinds of problems, they come with the capacity to be helpful in certain areas. Due to its power, from supposition and entanglement, quantum computers have great promise for many potential innovative regions as follows:

Artificial Intelligence: AI has become very powerful and promising today because it can use enormous data sets and learn from their experience. Machines take in data, process it in line with the expected result, continually receive feedback, and gradually become more accurate until they can act with a level of intelligence similar to humans.

The machine receives feedback by calculating the probabilities for many possible outcomes. The ability to receive feedback in this manner makes AI an ideal application for quantum computation.

Molecular Modelling: Following the quantum nature of chemical reactions, their properties are often highly entangled and super-positioned. Quantum computers, given their computational power, find it easier to evaluate these reactions. Consequently, they help create precision modelling of molecular interactions, identifying the ideal alignment for chemical reactions. With such precise molecular modelling, it becomes more efficient to make products such as solar cells, pharmaceutical drugs, and fertilisers.

Financial Modelling: Finance experts, stock market players, and investors constantly build models of real-world financial situations to forecast the performance of securities in the future.

However, there are no controlled settings to run financial modelling experiments in the stock market. Due to its stochastic nature, many financial operators or analysts look up to quantum computers to handle the complexity of the financial market. Quantum computing allows investors to evaluate the spread of possibilities or outcomes within numerous, randomly-generated scenarios.

Weather Forecasting: The ability to accurately predict weather greatly benefits many fields, including food production and transportation. That is besides the fact that it would help many people escape natural disasters such as hurricanes. However, accurately predicting weather requires processing vast data and many different variables. With its ability to manipulate complex data, quantum computing could significantly facilitate meteorologists' ability to predict the weather accurately.

Database searching: Quantum computers can comb through vast amounts of data and find patterns faster than classical or traditional computers.

12. Cloud Computing

Cloud computing involves delivering hosted services that offer computer system resources, including servers, databases, networking, computing power, and software over the Internet to store and retrieve data. With the help of cloud computing, users can store their data, files and even applications in remote servers that are very far away from where they reside; then, through the principle of on-demand, they can access their data or files (Microsoft Azure. n.d.). Many people today see cloud computing as synonymous with using the Internet. The origin of the cloud is traced back to when the Internet was depicted as a puffy cloud in the gigantic server-farm infrastructure, represented with flow charts and presentations in **Figure 18**.

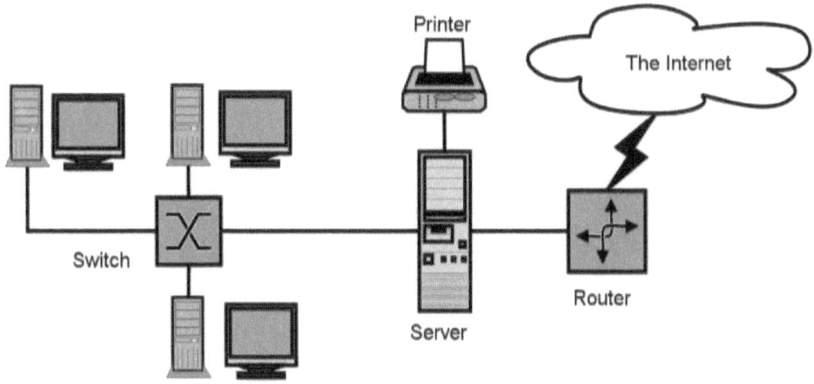

Figure 18: Gigantic Server-farm Infrastructure

When files and programmes are stored on the cloud, a user on the service can access them from anywhere. This automatically obviates the need to be close to any hardware. For instance, in the past, the way to access data or documents in one computer through another was to first save such document to a physical hard drive, USB drive or disk. Without such hardware, files originating on a particular computer were completely inaccessible on other computers. With cloud computing, documents and files are available. They can be easily accessed on any computer because they are embedded in a network of hosted servers that transmit data over the Internet.

How Cloud Computing Works

A front-end and a back-end process are attached to how cloud computing works. Those two ends are connected through a network, usually the Internet. The person using the computer to send files and applications to the cloud uses the front end, while the remote host system that stores what is sent uses the back end infrastructure.

On the front end is the customer's computer or computer network, as well as the applications necessary for accessing the cloud system. The back end comprises the systems, including computers, servers, storage devices and applications used at the other end to store all that is sent to them. Usually, each application in the cloud system has its dedicated server.

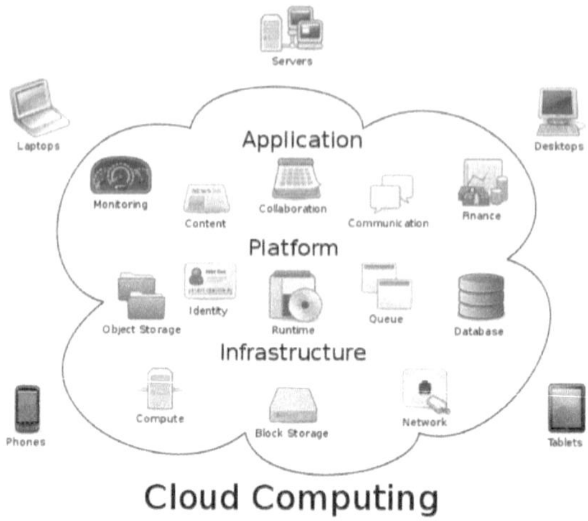

Figure 19: How Cloud Computing Works

Types of Cloud Computing Environments

A cloud environment refers to a collection of services offered to businesses to facilitate the effective functioning of their IT system. Usually, a cloud environment centralises an organisation's resources, enabling more efficiency in its operations. Proper set-up of the cloud computing environment facilitates remote workstations and remote working.

There are up to five types of cloud computing environments, also called cloud deployment. These are:

Public Cloud Environment: As the name suggests, this is a cloud environment in which several businesses share a cloud platform owned by an off-site expert to whom they pay a subscription. It is the most familiar cloud environment available, and most corporate organisations use it as their cloud environment of choice.

Private Cloud Environment: This is a cloud environment owned and used by only one organisation. Only the people working in the organisation can access the space and the files stored on it. A private cloud environment is usually situated on-site and managed by the user. It appeals primarily to businesses keen on exercising total control over the security of their files.

Hybrid Cloud Environment: This cloud environment is neither entirely public nor private. It combines the two types of environments. It is usually designed for businesses that require the features of both cloud environments for their cloud computing system. Businesses that adopt the hybrid cloud environment can take advantage of the public cloud environment's cost-saving and maintenance support and protect sensitive data through the platform's private backend.

In one hybrid cloud arrangement, there is a private cloud where data and proprietary applications are stored. However, whenever there is increased traffic, this model ports to a public cloud model to afford additional capacity for the private model. In the second type of hybrid cloud system, a private cloud contains most of the critical data and applications. In contrast, data that are neither critical nor important are sent to a public cloud system.

Community Cloud Environment: This cloud combines multiple public cloud services with a versatile, cost-efficient, and highly flexible private cloud environment. It is strongly advised that users use this type of cloud when more than one public cloud system is required for all the requirements of the user.

Cloud Computing Services

There are mainly four kinds of cloud computing that most users use. They are as follows:

Software-as-a-service (SaaS): This model gives customers a licence to use software applications on a pay-as-you-go or on-demand basis. These software applications, usually called web services, are delivered over the Internet so that customers or users can access SaaS applications and services from any location on any device with Internet access. Many organisations make use of software developed and stored by another organisation remotely. For example, many people today use Microsoft Office 365 remotely on a pay-as-you-go basis or on an on-demand basis. This is SaaS.

Infrastructure-as-a-service (IaaS): Through an outsourced, on-demand service, users procure virtual machines (VM) with various cloud resources, including operating systems, servers, software and APIs through IP-based connectivity. Users are allocated storage capacity on the infrastructure and can configure it with the other resources and use them based on their needs. An excellent example of IaaS is IBM Cloud and Microsoft Azure 2.3 systems.

Platform-as-a-service (PaaS): The PaaS system serves as a platform that allows users to develop software which the PaaS provider then hosts. Platforms such as Salesforce and Heroku4.5 host software development tools on their infrastructure. Users access these tools via APIs, web portals and software and leverage them to develop software hosted and delivered on the Internet by the PaaS provider. Lightning platform, Salesforce, Google App Engine, and AWS Elastic Beanstalk are well-known PaaS platforms.

Function-as-a-service (FaaS): FaaS is a concept based on serverless computing technologies and architecture that allows users to develop applications – particularly when building microservices applications by executing small, modular pieces of code known as functions – and deploy them in the cloud without maintaining a server. In FaaS, cloud service providers are responsible for the operation and maintenance of infrastructure. At the same time, developers of the software concentrate on building their applications, ensuring they are made to specifications, quality, agreed development timeline, cost and expected level of user-friendliness. Some notable implementations of FaaS are Open Whisk, Google Cloud Function, AWS Lambda, Microsoft Azure Functions, Webtask.io, IBM Cloud Functions, and Iron.io.

Benefits of Cloud Computing

Cloud computing comes with many benefits, which border on its flexibility, ability to work remotely, ability, and provision for disaster recovery.

It Offers Flexibility: Cloud computing offers flexibility that does not exist with traditional hosting. The flexibility allows users to quickly increase the volume of files and data on the cloud in real time without restarting servers. Users can also connect extra storage hardware or network bandwidth if there is a spike in traffic on their network. This helps to prevent crashes or downtime.

Allows Users to Work from Anywhere: Cloud computing allows any user on the network to access data and files via computers, smartphones, and other devices at any time. This means that people can work from anywhere and at any time, and everyone else on the network will instantly be aware of what every other person is doing or what they are working on.

Allows Automatic Updates: Cloud service companies must provide users timely updates about new versions of their systems, changes they make and all security issues that have arisen over time. This means that users can focus entirely on running the core aspects of their businesses without having to bother with software or security updates.

Helps with Disaster Recovery: It is inevitable that disasters such as network crashes, downtime and outages could occur in any business, leading to loss of data, productivity and revenue. Cloud computing can be very valuable in helping a business recover from any disaster that affects its operations; cloud computing services could recover whatever data may have been lost.

Cloud Services Offer High-Level Security: Cloud computing systems offer top-notch security that protects data against all malicious attacks. There are usually web application firewalls and advanced monitoring systems that assist in building a form of ring-fence around data to ensure its security. Moreover, cloud service providers always provide up-to-date encryption of all the data traffic from and to their systems, including in the storage of customer data. Such encryption makes data less accessible to hackers as well as people who do not have the right level of clearance or authorisation.

Future of Cloud Computing for Businesses

In 2019, more than 30% of decision-makers in IT enterprises considered the public cloud a top priority (Chai & Bigelow, 2021). However, public cloud adoption has yet to happen as predicted by experts, especially for mission-critical applications. This is expected to change in the coming years because business executives are increasingly demanding for public cloud to help their businesses compete globally.

Another advantage of the public cloud is its modernised internal systems, elasticity, and the empowerment of those crucial business units. Furthermore, cloud providers like VMware and IBM focus on satisfying business IT needs. This will aid the elimination of barriers confronting the adoption of the public cloud by decision-makers in IT.

Cloud architecture is gaining prominence as more businesses appear to be neglecting on-premises computer hardware for innovative solutions. Over the last couple of years, experts have predicted a rise in spending on the global cloud, with data stored in both public and private clouds expected to have a percentage increase. Customers will also likely demand that businesses fully adopt cloud computing. While the full manifestation of these predictions is expected, it is generally accepted that cloud computing will be here for some time.

13. Augmented Reality (AR)

AR is a technology and method of using a 3D virtual device to represent real-world phenomena and environments with the aid of an AR device, allowing the interaction of both the real-world and virtual objects to create the intended purpose (Hayes, 2014b). The overlaying of virtual objects and information in the real world enhances it with digital details like animation, images, and texts. These are viewed through AR screens, glasses, smartphones, or tablets (Harris, 2017; SmartTek, 2021).

Unlike virtual reality, where the real world is recreated and replaced with a virtual world, augmented reality focuses on enriching the idea of the natural world with digital images and information in the form of video, sound, infographics, photos, and others.

Types of Augmented Reality

AR consists of four types: Marker-based, Marker-less, Superimposition-based, and Projection-based.

Marker-based AR: This type of AR is also called a recognition-mode AR or image recognition AR. It relies on identifying markers-defined objects to function. It makes use of a marker to trigger augmentation. Markers consist of distinct patterns that enable cameras to recognise and process them. An example includes a marker-based mobile AR. The common feature is that they are visually independent of their environment.

The Marker-less AR: This is used in businesses, navigation apps, and events. This kind of technology, using location-based information, determines the type of content users find or get in specific areas. It uses tools such as compasses, accelerators, GPS, and gyroscopes.

Superimposition-based AR: This kind of AR partially or fully replaces an original object with an augmentation. The technology is mainly used by medical professionals, as seen in X-rays. It is also used for enhancing historical tours.

Project-based AR: This type of AR focuses on detecting the user's interaction with the surface using synthetic lights projected on surfaces. An example can be seen in holograms.

Benefits of Augmented Reality in Businesses

AR can monitor production and maintenance work, carry out a virtual walkthrough around properties or real estate, and improve remote designs and product advertisement, among others (GmbH, 2019; Okumoko, 2021; Porter & Heppelmann, 2017). Let us consider these benefits by industries.

In the manufacturing sector, project managers can remotely monitor work with the help of images with AR markers of manufacturing equipment. An example includes determining if a machine or device will fit into a location by pointing it to that location. Salespeople can create and publish fascinating AR branded content and add them to the ads so that people can engage with it to improve their knowledge of the product.

In retailing, it is natural for customers to "test" items before buying them, especially in fashion. Recently, AR has allowed retailers to try their online catalogues through smartphone screens.

In healthcare, surgeons can view critical digital images and information that can help them perform a surgical procedure. Medical students can use simulations of AR and VR to try out surgery procedures, with all the immersions and close-to-reality experience, reducing unnecessary risks and budgets.

In tourism, AR can assist travel agencies and brands in providing their potential clients with an immersive experience of a travel destination. They can provide their clients with image and video information of exciting places and signposts of their intended destination.

14. Virtual Reality (VR)

VR technology attempts to recreate computer videos and images that give real-life experiences visually that are entirely different from those seen on phones and computer screens. With the aid of a headset, VR applications completely immerse the user into a computer-replicated reality, providing realistic images and sounds and engaging all human sensory organs in an interactive and creative virtual world.

Unlike AR, VR requires that the user wear the display glasses or headset to have a 360 experience of the artificial world. Although VR is often associated with the entertainment and gaming industries, it has become a trend in other sectors such as construction, healthcare, the military, and engineering (Bernard Marr. 2021; Bousquet, 2018; Iberdrola, n.d.).

The Application of Virtual Reality

The applications of VR include the following:

Learning and training: VR is practical in a demonstration for trainees to get acquainted with the product or procedure of carrying out a process. The medical students mentioned can be seen as a practical example.

Work Collaboration: With VR, remote employee collaboration can be made possible. For instance, parties can work together on demo tasks that require visuals to enable them to understand and complete the task.

Managing Health-related Issues: VR has proven helpful in addressing health issues such as pain management, autism, and other social disorders. Depending on the situation, the technology helps the patient divert their brain from the pain, boost the brain's activity, or monitor the anxiety-causing symptoms to make them feel more comfortable or improve their recovery or social development.

Forecasting and Brainstorming: Businesses can virtually test their ideas before they are launched. Using VR, collaborators and partners can participate in the user experience, new models, and design testing phase of ideas.

The Future of Virtual Reality

As a very relevant technology, VR has a high possibility of expanding. It is fast becoming a crucial digital transformation framework for companies. Recently, the market demand for the technology has gone beyond the regular requirement for various disciplines of life to a more affordable application for users. Devices are currently in the market with powerful processors working effortlessly with Artificial Intelligence. As many companies continue to adopt this application, it is expected that in the nearest future, VR will shape the future and lead the advancement of industries and countries by extension.

15. Autonomous Vehicles

Driverless vehicles can operate without the involvement of humans and through their capacity to sense their environment. These vehicles can do everything and go everywhere a regular car can (Schaub & West, n.d.; Sleight, 2021).

There are different categories of vehicle driving automation. They start from level 0 (where there is hardly any automation), through levels 2, 3, and 4, to level 5 (where there is total automation and the vehicle can autonomously drive itself). They are as presented in **Figure 20** (Synopsys, n.d.):

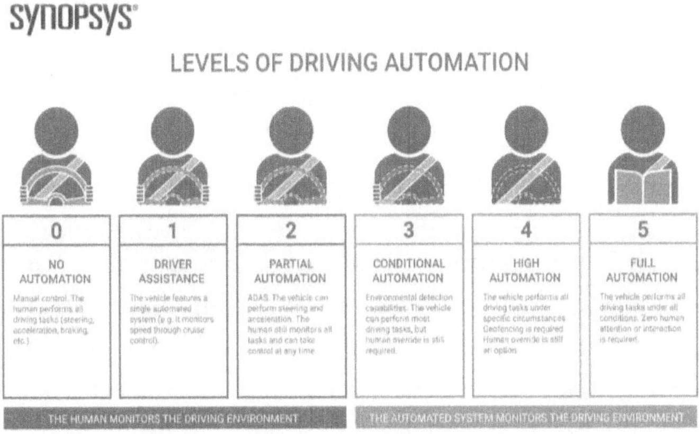

Figure 20: Levels of Driving Automation

Level 0: A human is in charge of driving, as the vehicle does not have any automated contribution to aid the physical driving of the car.

Level 1: A human is assisted by braking and accelerating or steering mechanism using an advanced driver assistance system (ADAS).

Level 2: At this level, a human carries out most of the physical driving of the vehicle, besides the vehicle's ability to assist with automatic braking, acceleration and steering.

Level 3: Here, the vehicle, through its automated driving assistance system (ADAS) and control by a human driver, can handle all the driving functions. However, when the human wants to take back control of the vehicle, or the ADAS instructs him to do so, he must be able to take control of driving the car immediately.

Level 4: At this level, the ADAS can carry out all the driving functions and continually monitor the environment without the need for vigilance on the part of a human driver.

Level 5: At this level, the vehicle is fully automated and can drive itself autonomously without the intervention or attention of an accompanying human.

How Autonomous Vehicles Work

Autonomous vehicles, as they are often referred to, rely on complex algorithms, actuators, sensors, powerful processors, and machine learning systems to execute their software. With the help of the sensors located in several parts of their body, these cars develop a sensory ability to map out their surroundings. The radar sensors and video cameras help it track and locate nearby cars, detect road signs and traffic lights, and watch out for pedestrians (Sleight, 2021).

The light detection and ranging (LIDAR) sensors exchange light pulses with the surrounding vehicles to calculate distances, identify markings and lanes, and detect road edges. When parking, ultrasonic sensors built into the wheels place curbs and the cars around.

The sophisticated software processes all these sensory inputs, plot its path and give directions to the vehicle's actuators. Once this happens, other car functions, such as acceleration, steering, and braking, are controlled. Other systems, such as predictive modelling, hard-coded rules, object recognition, and barrier avoidance algorithms, assist the software in navigating through obstacles and obeying traffic rules.

Benefits of Autonomous Vehicles

Autonomous vehicles provide several benefits to the countries that have adopted them. One of such is traffic fatalities. These have been significantly reduced as these vehicles have better precision and fewer human errors. It also promotes relaxation and privacy when commuting, especially among workers and travellers. Mobility is also a benefit where the vehicles can swiftly move and access lifesaving facilities or services. The technology also supports the physically challenged in managing their movement and gaining direct access to healthcare services.

Challenges of Autonomous Vehicles

The challenges of this kind of technology range from legislative and technological to philosophical and environmental issues. The LIDAR is quite expensive, and efforts are still being made to strike the appropriate balance between resolution and range. Unstable weather conditions are another issue, as the cameras and sensors may have problems working properly in extreme weather conditions. There are also questions surrounding the performance of autonomous vehicles in adhering to traffic laws and conditions. Examples of such inquiries are if these vehicles would be given a specific lane and how they would fare in gridlocks. Then, there is the emotional and artificial intelligence of these vehicles, which relates to how they can read body language, make eye contact, make instant decisions, predict behaviours, or make split-second judgments. Other notable challenges in environments like the US are the federal and state regulations and the liability of accidents.

16. Smart Cities

Smart cities have a functional technological structure that uses the Internet of Things (IoT) to collect and analyse data for the operation and management of the city (Woherem & Odedra-Straub, 2017). The IoT tools include connected lights, sensors, and meters. The data are used to improve public infrastructures, utilities, and others. The idea of smart cities is growing as urbanisation over the years has required cities to find alternative ways to manage resources while developing their economies (Business Insider, 2022; Nexus Integra EN, 2021). The aim is to improve a population's quality of life by optimising the cities' functions and promoting network advancement through smart technologies and data analysis.

Technologies for a Smart City

The following technologies make it easier for a smart city to run smoothly:

Smart Utility Metres: A smart metre is a device that makes daily tasks and operations efficient and more accessible while promoting public safety and reducing traffic and other environmental issues.

Smart Grid: In some quarters, a smart grid is regarded as the most excellent way to accomplish a smart city because it tremendously helps with conserving resources. Cities like Amsterdam experimented with providing storage units for home energy and solar panels for houses directly connected to the grid of the smart city.

Smart Solutions for Waste Management: Managing waste can be costly and can easily expose inefficiency when they build up. The solution smart waste management offers helps to eliminate these challenges by monitoring trash cans and feeding the data to the management companies, who will then provide a pick-up service.

Smart Monitor of Air Quality: The system detects particles, dirt, specks of dust, and cleaning chemicals that hover around the airspace in homes or offices and informs the pollutant users.

References

Allen, R. (n.d.). *Types of big data-understanding & interacting with key types.* SelectHub. Retrieved May 10, 2022, from https://www.selecthub.com/big-data-analytics/types-of-big-data-analytics/

Bernard Marr. (2021, July 2). *The 10+ best real-world examples of augmented reality.* Bernard Marr. Retrieved May 10, 2022, from https://bernardmarr.com/the-10-best-real-world-examples-of-augmented-reality/

Bousquet, C. (2018, September 24).*9 potential applications for AR and VR technology.* Data-Smart City Solutions. Retrieved May 10, 2022, from https://datasmart.ash.harvard.edu/node/1406484.json

Business Insider. (2022, February 5). *IoT and Smart City Technology: How connected cities work.* Business Insider. Retrieved May 10, 2022, from https://www.businessinsider.com/iot-smart-city-technology?r=US&IR=T

CB Insights Research. (2021). *What is quantum computing? Definition, industry trends & benefits explained.* CB Insights Research. Retrieved May 10, 2022, from https://www.cbinsights.com/research/report/quantum-computing/

Chai, W., & Bigelow, S. J. (2021, December 16). *What is cloud computing? Everything you need to know.* TechTarget. https://www.techtarget.com/

searchcloudcomputing/definition/cloud-computing

Cisco. (2018). *Cisco Annual Internet Report - Cisco AIR Infographic (2018 - 2023).* Cisco. Retrieved May 19, 2022, from https://www.cisco.com/c/en/us/solutions/executive-perspectives/annual-Internet-report/infographic-c82-741491.html

Cossins, D. (2019, October 23). *It's official: Google has achieved quantum supremacy.* New Scientist. https://www.newscientist.com/article/2220968-its-official-google-has-achieved-quantum-supremacy/

Daley, S. (2021, December 21). *34 Blockchain applications and real-world use cases disrupting the status quo.* Built In. Retrieved May 9, 2022, from https://builtin.com/blockchain/blockchain-applications

Edureka. (2020, November 25). *IoT applications, Internet of things examples, Real world IoT examples.* Edureka. Retrieved May 9, 2022, from https://www.edureka.co/blog/iot-applications/

EMF. (n.d.). *5G Explained - How 5G works.* EMF Explained. Retrieved May 10, 2022, from http://www.emfexplained.info/?ID=25916

Frankenfield, J. (2018, May 26). *Quantum computing definition. Investopedia.* Investopedia. Retrieved May 10, 2022, from https://www.investopedia.com/terms/q/quantum-computing.asp

FutureLearn. (2021, October 15). *What is quantum computing? Essential concepts and uses.* FutureLearn. Retrieved May 10, 2022, from https://www.futurelearn.com/info/blog/what-is-quantum-computing

Futurum Research. (2022). *Global smart speaker market share loup ventures.* Futurum Research. Retrieved May 19, 2022, from https://futurumresearch.com/smart-homes-smart-speakers-and-the-Internet-of-google/global-smart-speaker-market-share-loup-ventures-01/

GE Additive. (n.d.). *What is Additive Manufacturing? Building a world that works.* General Electric. Retrieved May 10, 2022, from https://www.ge.com/additive/additive-manufacturing

Gillis, A. S., & Gerwig, K. (2022, March 15). *What is 5G? Definition, benefits and use cases.* TechTarget. Retrieved May 10, 2022, from https://www.techtarget.com/searchnetworking/definition/5G

GmbH, D. (2019, August 8). *Application areas of Augmented Reality.* GAL Digital. Retrieved May 10, 2022, from https://www.gal-digital.de/en/digitalization-blog/application-augmented-reality

Hagel, J., Brown, J.S., Samoylova, T. and Liu, M. (2013). *From exponential technologies to exponential innovation. Report 2 of the 2013 Shift Index series.* Deloitte Insights. Retrieved May 9, 2022 from: https://www2.deloitte.com/insights/us/en/industry/technology/from-exponential-technologies-to-exponential-innovation.html.

Harris, M. (2017, August 10). *Everything you need to know about designing with augmented reality.* Digital Arts Online. Retrieved May 10, 2022, from https://www.digitalartsonline.co.uk/features/hacking-maker/everything-you-need-know-about-designing-with-augmented-reality/

Harris, T., & Pollette, C. (2002, April 16). *Robot basics.* HowStuffWorks. https://science.howstuffworks.com/robot1.htm

Hayes, A. (2014a, August 28). *3D printing definition.* Investopedia. https://www.investopedia.com/terms/1/3d-printing.asp

Hayes, A. (2014b, September 15). *Augmented Reality (AR) definition.* Investopedia. https://www.investopedia.com/terms/a/augmented-reality.asp

Iberdrola. (n.d.). *Virtual Reality, the technology of the future.* Iberdrola. Retrieved May 10, 2022, from https://www.iberdrola.com/innovation/virtual-reality

Intel. (n.d.). *A guide to the Internet of Things.* Intel. Retrieved May 19, 2022, from https://www.intel.co.uk/content/www/uk/en/Internet-of-things/infographics/guide-to-iot-new.html

Juniper Research. (2020). *"Internet of Things" connected devices to triple by 2021.* Juniper Research. Retrieved May 19, 2022, from https://www.juniperresearch.com/press/Internet-of-things-connected-devices-triple-2021

Kaspersky. (2022, March 30). *What is cyber security? Definition, types, and user protection.* Kaspersky. Retrieved May 10, 2022, from https://www.kaspersky.com/resource-center/definitions/what-is-cyber-security

Levy, A. (2021, October 25). *15 applications for blockchain technology.* The Motley Fool. Retrieved May 10, 2022, from https://www.fool.com/investing/stock-market/market-sectors/financials/blockchain-stocks/blockchain-applications/

Martech Advisor. (2019, March 4). *By 2030, each person will own 15 connected devices-here's what that means for your business and content.* Martech Advisor. Retrieved May 19, 2022, from https://www.martechadvisor.com/articles/iot/by-2030-each-person-will-own-15-connected-devices-heres-what-that-means-for-your-business-and-content/

Matob R. (2021, December 4). *Internet of things (IoT): Definition, how it works, example and benefits.* Matob. Retrieved May 9, 2022, from https://matob.web.id/random/technology/Internet-of-things/

McGeoch, C. (n.d.). *Experimental evaluation of an adiabatic quantum computer for combinatorial optimization.* Computer Science. Retrieved May 19,

2022, from https://csci.williams.edu/catherine-mcgeoch/

Methods and goals in AI. (2022, March 18). Encyclopaedia Britannica. https://www.britannica.com/technology/artificial-intelligence/Methods-and-goals-in-AI

Microsoft Azure. (n.d.). *What is cloud computing? A beginner's guide. Microsoft Azure, Cloud Computing Services*. Retrieved May 10, 2022, from https://azure.microsoft.com/en-us/overview/what-is-cloud-computing/benefits

Muthoni, J. (2021, June 16). *Use of robotics in our daily lives.* JonasMuthoni.com. Retrieved May 10, 2022, from https://jonasmuthoni.com/blog/use-of-robotics-in-daily-lives/

Nexus Integra EN. (2021, October 1). *Smart Cities, What are and how do they work?* Nexus Integra. Retrieved May 10, 2022, from https://nexusintegra.io/smart-city-what-is-how-it-works/

Okumoko, J. (2021, May 24). *10 Cool Applications of AR Technology in Everyday Life*. MUO. Retrieved May 10, 2022, from https://www.makeuseof.com/cool-applications-of-ar-technology/

Popova, M. (2012, October 11). *As We May Think: Vannevar Bush's Prescient 1945 Vision for the Information Age, the Power of "Curation," and the Need for Open-Access Science.* The Marginalian. https://www.themarginalian.org/2012/10/11/as-we-may-think-1945/

Porter, M. E. & Heppelmann, J. E. (2017, November 1). *How does augmented reality work?* Harvard Business Review. Retrieved May 10, 2022, from https://hbr.org/2017/11/how-does-augmented-reality-work

Qualcomm. (2017, July 25). *What is 5G? Everything you need to know about 5G.* Qualcomm. Retrieved May 10, 2022, from https://www.qualcomm.com/5g/what-is-5g

Remmert, H. (2019, November 25). *5G applications and use cases.* Digi International. Retrieved May 10, 2022, from https://www.digi.com/blog/post/5g-applications-and-use-cases

Rouse, M. (2011, June 14). *What is cybersecurity?* Techopedia. https://www.techopedia.com/definition/24747/cybersecurity

Schaub, H., & West, D. M. (n.d.). What is a driverless car? *Brookings*. Retrieved May 10, 2022, from https://www.brookings.edu/blog/techtank/2015/05/21/what-is-a-driverless-car/

Security Today. (2020, January 13). *The IoT rundown for 2020: Stats, risks, and solutions.* Security Today. Retrieved May 19, 2022, from https://securitytoday.com/Articles/2020/01/13/The-IoT-Rundown-for-2020.aspx?Page=2:~:text=During%202020%E2%80%94experts%20estimate%20the,be%20connected%20to%20the%20web

Simplilearn. (2020, May 11). *What is blockchain technology, and how does it* *work?*Simplilearn.Com. Retrieved May 9, 2022, from https://www.simplilearn.com/tutorials/blockchain-tutorial/blockchain-technology

Sleight, M. (2021, June 22). *How do self-driving cars work?* Bankrate.com. Retrieved May 10, 2022, from https://www.bankrate.com/insurance/car/how-do-self-driving-cars-work/

SmartTek. (2021, September 2). *What is Augmented Reality (AR), and how does it work?*SmartTekas.com. Retrieved May 10, 2022, from https://smartteksas.com/blog/what-is-augmented-reality-ar-and-how-does-it-work/

Southern Illinois University Carbondale. (2016). A brief history of IT, Computer Technical Support Newsletter, 2 (29), 1-6.Retrieved May 9, 2022, from https://ehs.siu.edu/_common/documents/IT%20newsletter/vol-2-no-29.pdf

Spatial Team. (2020, July 6). *The Additive Manufacturing Process*. Dassault Systèmes, Spatial Corp. Retrieved May 10, 2022, from https://blog.spatial.com/additive-manufacturing-process.

Statista. (2022, February 1). *IoT and analytics global revenue by segment 2021.* Statista. Retrieved May 19, 2022, from https://www.statista.com/statistics/913299/projected-global-revenue-of-the-Internet-of-things-segment/

Stezano, M. (2018, September 12). *In 1950, Alan Turing Created a Chess Computer Program That Prefigured A.I.* History Stories. Retrieved May 9, 2022, from https://www.history.com/news/in-1950-alan-turing-created-a-chess-computer-program-that-prefigured-a-i

Synopsys. (n.d.). *What is an autonomous car? – How self-driving cars work.* Synopsys. Retrieved May 10, 2022, from https://www.synopsys.com/automotive/what-is-autonomous-car.html

Taylor, D. (2020, January 6). *What is big data? Introduction, types, characteristics.* Guru99.com. Retrieved May 10, 2022, from https://www.guru99.com/what-is-big-data.html

Tech 21 Century. (2020, August 3). *18 benefits of the Internet of things (IoT) in our lives and businesses.* Tech21Century. Retrieved May 9, 2022, from https://www.tech21century.com/Internet-of-things-iot-benefits/

Travaly, Y., & Muvunyi, K. (2020, January 13). *The future is intelligent: Harnessing the potential of artificial intelligence in Africa.* Brookings. https://www.brookings.edu/blog/africa-in-focus/2020/01/13/the-future-is-intelligent-harnessing-the-potential-of-artificial-intelligence-in-africa/

UpGrad Blog. (2020, May 6). *What is Big Data - Characteristics, types, benefits*

& examples? upGrad Education Private Limited. Retrieved May 10, 2022, fromhttps://www.upgrad.com/blog/what-is-big-data-types-characteristics-benefits-and-examples/

Vogel, S. (2019, November 6). *What are the benefits of quantum computing?* IT PRO. Retrieved May 10, 2022, from https://www.itpro.co.uk/technology/34752/what-are-the-benefits-of-quantum-computing

Wikipedia contributors. (2021, December 22). *Computing Machinery and Intelligence.* Wikipedia. Retrieved 12:36, May 9, 2022, from https://en.wikipedia.org/w/index.php?title=Computing_Machinery_and_Intelligence&oldid=1061598194

Woherem, E. E., & Odedra - Straub, M. (2017). The potentials and challenges of developing smart cities in Africa. *Circulation in Computer Science, 2*(4), 27-39.

Woherem, E. E. (1993). *Information Technology in Africa: Challenges and Opportunities.* African Centre for Technology Studies (ACTS) and Research Institute for Knowledge Systems.

Zero Hour Parts. (2020, July 13). What is additive manufacturing? - Defining Additive manufacturing process. *Zero Hour Parts.* Retrieved May 10, 2022, from https://zerohourparts.com/what-is-additive-manufacturing-processes/

Chapter 6
Artificial Intelligence and the Fourth Industrial Revolution

The emergence of exponential technologies such as artificial intelligence (AI), the Internet of Things (IoT), big data, quantum computing, the blockchain, nanotechnology, VR and AR, and others have already begun to change production and the economies of the countries of the world dramatically.

The World Economic Forum's executive chairman and founder, Klaus Schwab, referred to this new age as "The Fourth Industrial Revolution" (4IR). This is a new frontier of the digital industrial revolution. According to Schwab, it is marked by the blurring of many things that are "physical, digital, and biological" in their different spheres.

While there is a consensus among scholars and scientists that all the exponential technologies will have a tremendous impact on production, economies, and people's lives generally, most scientists attribute greater importance to artificial intelligence.

The Chair of Artificial Intelligence and Robotics at Singularity University, Neil Jacobstein, claims that "AI is not simply a game-changer; it will disrupt the whole playing field" (Singularity, 2017). Similarly, Jason Silva, a futurist and techno-philosopher, stated that artificial intelligence is likely the parent of all exponential technologies. Indeed, it brings about global and human-race changes that we can barely comprehend (Singularity, 2017).

Andrew Ng, a computer scientist, co-founder of Coursera and an adjunct professor at Stanford University, esteems artificial intelligence higher than other exponential technologies, describing it as "the new electricity." In his words, "AI is the new electricity. I find it hard to envisage a part of the economy that AI will not change. The widespread use of AI to propel technological and business ventures is a path to an AI-powered future" (Jewell, 2019).

The coming of the steam engine gave rise to "the first industrial revolution." In 1834, the electric motor was invented, ushering in the second industrial revolution, where steam-based machines were displaced with electric-based machines in mechanical processes. In the third industrial revolution, computers and the Internet were the game changers. As electricity became the source of power during the second industrial revolution and has since become the mainstay of modern life, AI will power all other exponential technologies during the fourth industrial revolution. It will crawl into everything, just as electricity has done in the past 100 years. Thus, practically every aspect of modern human life and endeavours - in the industrial sectors, agriculture, education, government, offices, and homes - will be integrated with AI. It will become an embedded technology found in virtually every product we use without even being noticed or bothered about.

What Is Artificial Intelligence?

The previous chapter explained artificial intelligence as a technology that seeks to design computer systems that simulate human intelligence or cognitive ability. Thus, computers embedded with AI can execute tasks intelligently and accurately.

Types of Artificial Intelligence

There are two major types of AI. They are now referred to as Type-1 and Type-2 AI systems, as illustrated below.

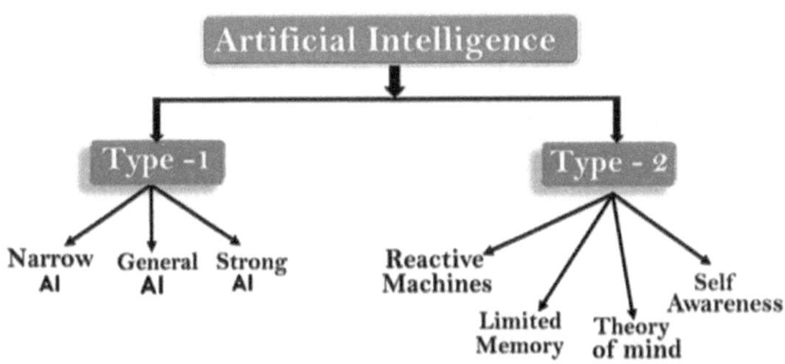

There are two bases for classifying artificial intelligence: capabilities and functionality.

Type 1 AI – Based on Capabilities

i. Weak AI or Narrow AI: This is today's most familiar form of AI. These AI systems have limited capabilities because they perform one particular task or solve a specific problem. Usually, they execute the task well, sometimes surpassing human functioning in the specific context.

An excellent example of Narrow AI is Apple's Siri. Other examples of the employment of Narrow AI include purchasing suggestions on e-commerce sites, self-driving cars, speech recognition, and image recognition.

ii. General AI: The idea behind General AI is to build computer systems that will mimic or even be defined as possessing general human cognitive ability across various domains, including reasoning, problem-solving, decision-making, language understanding and speaking, learning, exhibiting common sense knowledge, and game playing.

There are many instances of using Narrow AI systems today, but there is no case of an acceptably fully developed General AI system. Having said that, in many laboratories and departments of the world, cutting-edge research is being carried out to create a general AI system. Therefore, many people will argue that General AI is still a theoretical concept rather than an actualised machine or software system. Building a machine with the capabilities of Artificial General Intelligence would require the aggregation of thousands of Narrow AI systems working in sync to simulate human intelligence fully.

Nevertheless, several computer scientists are actively involved in AGI research with the hope of fully replicating human intelligence in machines soon. An AGI will pass the Turing Test.

iii. Super AI (Strong AI): This is a level of Artificial Super Intelligence (ASI) at which machines will not just mimic human intelligence but will exhibit capabilities that will supersede them. These computer systems will show capabilities in rational decision-making, solving puzzles, making judgments, planning, communicating independently, and building emotional relationships.

Scientists expect that ASI will be a natural progression from Artificial General Intelligence.

Once computer scientists achieve AGI, it will take relatively less time and effort to advance AI systems into the realm of ASI.

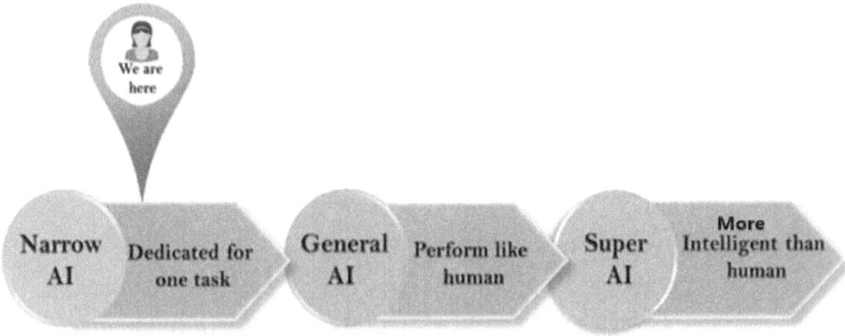

Figure 21:***Projection of Type 1 AI: from Narrow AI to Super AI***

Type 2 AI – Based on Functionality

i. Reactive Machines: This AI is designed to perform basic operations. They are programmed to execute a specific task, leading to a predictable output based on their input. They cannot learn actions and do not store memories or past experiences to determine future actions. AlphaGo and IBM's Deep Blue are good examples of Reactive AI systems. Email spam checkers, supercomputers that play chess, and a Netflix recommendation engine are also examples.

ii. Limited Memory: Limited Memory AI has become today's most widely used AI. These AI systems can store past experiences or previous data for a limited period. They can then use the data from their previous experiences and other current information to predict the future or solve complex problems.

Autonomous vehicles are germane examples of systems that are imbued with limited memory. They are cognisant and have the memory of the speed every other car on the same road is driving, how far or near the vehicles are, the direction they are heading towards and the weather condition on the road. They use this stored history and pre-programmed information to seamlessly navigate the road, going from one place to another.

However, as their name suggests, Limited Machine AI is limited. The information stored in its memory lasts for only a short period.

iii. Theory of Mind AI: Presently, these computer systems are not in existence, but computer scientists and researchers are working hard and making much effort to develop them. The intention is to create machines that understand and exhibit the different range of human intentionality, ranging from knowledge, communication, speechwriting, emotions and beliefs, and become capable of social interaction, holding intelligent conversations like humans.

As these machines can understand human emotions and feelings, they can adjust their behaviour to suit each unique circumstance as they interact with people. They will also possess decision-making capabilities that will be the same as humans.

The above features of the Theory of Mind AI are impressive. It is important to note that there is still much work to achieve a quantum leap in society's productivity. For example, it is a fact that every organisation has employees, and each employee brings with them, their beliefs, emotions, dreams, and apprehensions to the workplace. Simply put, they bring their minds to work. A worker's mental state significantly impacts their ability to be productive at work, which is why enterprises should positively leverage their employees' emotions to increase productivity. For enterprises to prosper, AI must build such "team members" that extend far beyond chatbots.

AI can only achieve this if the artificial intelligence-based "team members" comprehend the feelings of their human teammates, beliefs, and objectives, among other things. This is only conceivable if the AI-powered teammates have a theory of mind. This is the primary condition that Theory of Mind AI aims to meet. Meanwhile, Artificial Emotional Intelligence and the Theory of Decision-Making are two AI domains leading research toward building the Theory of Mind AI.

iv. Self-Awareness – Like the Theory of Mind AI, Self-Aware AI is also the future of AI. These systems will be more intelligent than humans, possessing emotions, consciousness, and sentiments. Like humans, they will be fully self-aware of their needs and desires.

The self-aware nature of these machines means that they will be aware of their feelings and those of the people around them. They will also be mindful of their mental states and will be able to make inferences.

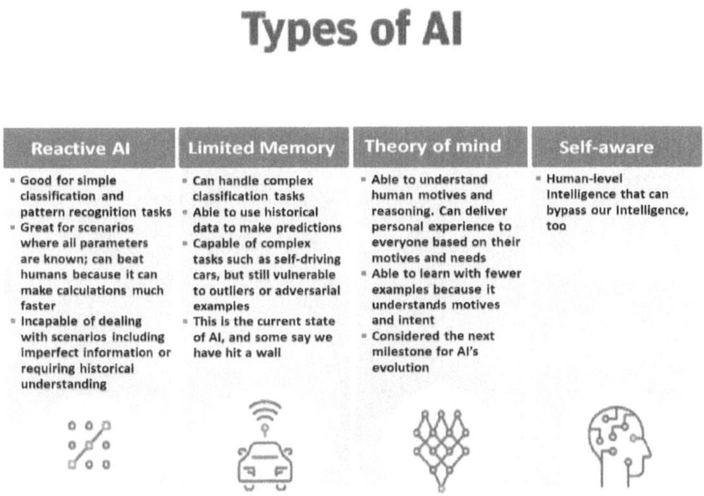

Figure 22: Types of AI
Source: Adapted from TechTarget.

How Does AI Work?

AI systems work intelligently as they have enormous amounts of data; with techniques to analyse and make quick decisions based on intelligent algorithms and patterns found in the data, they can establish a deep understanding of situations and make quick decisions. As a discipline, AI has various sub-domains which individuals can utilise in multiple fields and industries. The key areas of AI are as follows:

1. Machine Learning: This is an aspect of AI that deals with the ability of machines to learn from past experiences, adjust to new data inputs, identify patterns, perform human-like tasks and make decisions with little or no human intervention. AI generally seeks to build machines that mimic human cognitive abilities; thus, machine learning is a specific domain of AI that trains machines to learn. Product recommendations on shopping websites, spam filters in emails, and chatbots are all powered by machine learning.

2. Deep Learning: Deep learning is a sub-field of machine learning that involves algorithms called neural networks, modelled after the structure and function of the human brain. Through deep learning, computer systems learn to process data via layers to classify, identify patterns, infer, and predict outcomes. Deep learning involves functioning driverless vehicles, virtual assistants, chatbots, medical research, and facial recognition.

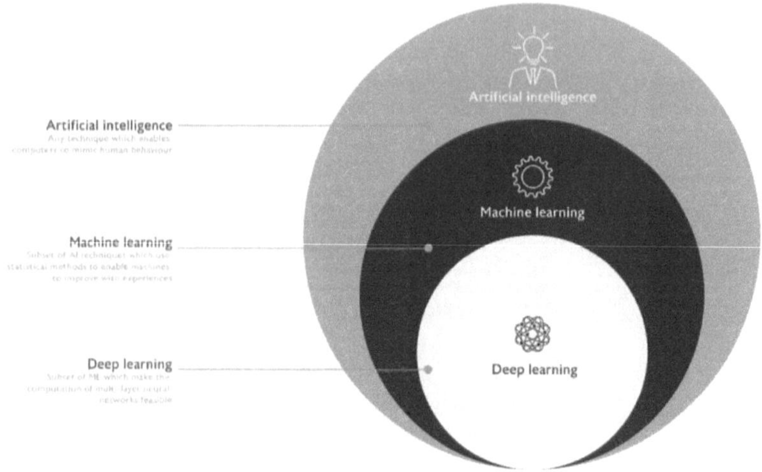

Figure 23: AI, Machine Learning and Deep Learning

3. Neural Networks: They are primarily used in deep learning and are a network of algorithms that work on a similar principle to neurons in the human brain. They analyse data to determine the relationships and patterns in the variables and generate an output that may or may not go to the next layer of the network, depending on whether the result generated by any individual node is above a specified threshold.

The concept of neural networks is useful in various situations, including developing trading systems for the finance industry, security classification, and credit risk modelling.

4. Natural language Processing, NLP: This is an aspect of AI concerned with building computer systems that process human language in the form of texts and spoken words and understand the intent of the writer or speaker, just the same way humans do. They read texts, understand spoken words, respond to texts or spoken words, and convert text and voice from one language to another. The use of smart assistants (Siri, Alexa, and Cortana), text analytics, email filters, digital phone conversations, language translation, and many other applications of NLP technology are only a few examples of the many real-world applications of NLP technology that have a significant impact on our day-to-day lives.

5. Computer Vision: Computer vision is an aspect of AI that focuses on training computers to "see," observe, understand, and accurately interpret the world around them using cameras, data, and algorithms. This is carried out in a manner akin to human vision. It enables computer systems to evaluate digital images, videos, and other visual data to determine one from another. Also, it allows computers to identify various features of the visual data fed into them - such as how far away they are, whether they are moving or stationary, and whether a particular object is wholesome or defective. Some computer vision applications include detecting metal cracks, paint defects, bad prints, assembly verification, screen reading for extracting data, code and character (typology) reading, and intruder detection.

6. Cognitive Computing: This is a subdivision of AI whereby computer systems try to mimic the human thought process, making decision-making more innovative and more accessible and using self-learning algorithms; these systems leverage data mining, pattern recognition, and natural language processing to process information by comparing it with a teaching set of valuables to generate an outcome. Areas of application of cognitive computing include enabling the blind to navigate a physical space, helping investment strategists to make buying and selling decisions, and providing more insightful knowledge of diseases and courses of treatment.

Factors Enhancing the Dominance of AI

Although computer scientists have explored the possibility of building AI-powered computer systems that will drive the world, they did not make much progress given the technology that existed at the time. However, over the past several years, many factors have emerged to fast-track the development and adoption of AI. These factors include the following:

i. The cost of computer processing power has fallen while the number of chips incorporated in computers has increased drastically over the past several years. The technology for chip development is rapidly evolving, producing new chips with more excellent functionalities, speed and processing power every year.

ii. The cost of gigabytes of storage and cloud access has fallen, creating much more storage power. Meanwhile, the availability of data has exponentially increased.

iii. There has been an increased understanding of neural science, specifically the human brain. As we know more about cognitive science and the brain's functioning, knowledge is used to develop even more sophisticated and intelligent AI systems.

Besides the convergence of the factors listed here, the ability of humans to learn and apply AI algorithms has also grown. As the ability to build more applications increases, AI will become even more dominant across the various fields of human endeavour.

The Potential Economic Impact of AI

Like electricity as an energy source during the second industrial revolution, AI tremendously impacts the production process and the economies of corporations and nations during the fourth industrial revolution. Studies undertaken by various bodies, organisations, and institutions worldwide have predicted this much. A 2017 research conducted by PwC (2018) estimated that during the decade between 2017 and 2027, the Chinese economy would experience an increase of 26% in GDP due to the country's heavy investment in AI. The report also estimated that by 2030, global GDP could increase by as much as 14% due to the injection of $15.7 trillion from AI-powered economic activities.

Meanwhile, a 2018 research conducted by PwC concluded that the global market value for AI will hit $169 billion by 2025, while AI algorithms will power $3.9 trillion in global business value by the same year. PwC also opines that by 2030 AI's contribution to the global economy will increase to $15.7 trillion. This is mind-boggling. According to the report, practically every sector of the world's economy will benefit from AI, which is estimated to be immensely beneficial for all industries. However, the service industry will be the most positively impacted sector, particularly health, education, and public services, which will grow by 21%. Despite the importance of the financial services sector, AI will make a minimal contribution of 10% to the industry.

According to Accenture (2016), AI will increase by twofold the economic growth rate in some key developed countries by 2035.

Of the 12 economies, the United States will experience the highest economic benefit, yielding a 4.6% growth rate by the same year. These economies collectively generate more than 50% of the global economic output. While these economic changes would be ongoing, AI will increase labour productivity in these countries by 40%.

In their report of 2018, McKinsey & Company predicted that AI could potentially add around $13 trillion to the global economy. It is estimated that about 70% of all the companies in the world will make use of at least one AI technology in their operation and that less than 50% of big companies will make use of all types of AI tools by the year 2030 (Bughin, Seong, Manyika, Chui & Joshi, 2018).

All the research and studies highlighted above predict massive adoption of AI technologies by corporations which will significantly boost global GDP by 2030, leading to a more prosperous society.

AI and the Fourth Industrial Revolution

"With automation, there will come abundance. Almost everything will get very cheap." - Elon Musk.

While the application of AI across various domains has become more dominant, it is clear that the accomplishments recorded so far are merely the tip of the iceberg compared to what future AI breakthroughs will bring.

As the fourth industrial revolution gets fully underway, with the amalgamation of AI and automated machines, the resultant effect will dramatically transform the world, affecting human labour, improving the quality of life, empowering governments, and raising ethical issues.

Consolidating automated machines with intelligent computer systems has birthed machines that conquer complex tasks more efficiently and faster than humans. Robots are more efficient than humans in many difficult jobs in many sectors, such as logistics, medicine, manufacturing, and hospitality. In many global companies, AI-powered chatbots have become quite prolific, with experts estimating that all customer interactions will be with chatbots by 2025. In the future, chatbots will most likely take over the job of call centre agents. They will also perform tasks such as interviewing job candidates, delivering training and handling in-store sales.

In his book, *You and AI: A Citizen's Guide to AI, Blockchain, and Puzzling Together the Future of Healthcare*, Ronanki (2021) envisions a future in which the mass adoption of AI, blockchain, IoT, big data, and other exponential technologies by businesses, governments, and specialised services will transform society for the better. Moreover, Ronanki (2021) predicted that the healthcare sector would likely experience the most significant impact of adopting exponential technologies. AI can provide proactive, predictive, and personalised healthcare at a lower cost for all.

Will AI Take Over People's Jobs?

As it has been apparent so far, AI technologies have brought many benefits and will get even more in the years ahead as the application of AI permeates more domains of human endeavour. Nevertheless, many people still fear AI would negatively affect human existence.

Unfortunately, the media and Hollywood have not helped to assuage these fears. Many reports in the media have zeroed in and overly highlighted the potentially harmful effects of AI (Emen, 2020). Hollywood is full of sci-fi movies that depict AI-powered robots overpowering humanity, causing unprecedented havoc and sometimes even wiping out portions of the earth or humanity.

AI Will Replace Some Jobs

While there are no clear indications that the human race is at risk of attacks from AI soon, there is a negative impact of AI that may confront humanity sooner than later: unemployment.

As AI-powered automation becomes more ubiquitous, many experts believe machines will take over many professional jobs by 2030. According to the Forbes Technology Council (2019), the jobs affected are "insurance underwriting, warehouse and manufacturing, research and data entry, long haul trucking, production line, and local TV advertising." Others include pharmaceutical discovery, bank tellers and cashiers, outbound sales, fast food service, package delivery services, and "Any Tasks That Can Be Learned."

In his 2018 book, *AI Superpowers: China, Silicon Valley, and the New World Order*, AI guru Kai-Fu Lee asserts that AI would automate "50% of all jobs within 15 years." According to him, "accountants, factory workers, truckers, paralegals, and radiologists - just to name a few - will be confronted by a disruption akin to that faced by farmers during the Industrial Revolution" (Lee, 2018).

AI has already started to shift employee engagement levels across the world. AI-powered robots are already decimating blue-collar jobs. Greatly to be affected are factory workers, cooks, messengers, servers, gardeners, cleaners, drivers, and construction workers. As AI-powered systems become more efficient and ubiquitous, they will take over much of the work currently done by medical professionals, salespeople and accountants, delivery truck drivers, and others.

Incorporating AI into production will inevitably lead to a loss of jobs. According to a study by PwC (2017), 30% of employment in the UK could be at risk due to automation by 2030. For the USA, Germany, and Japan, the percentage of jobs that could be at risk due to automation is 38%, 35%, and 21%, respectively.

In terms of the respective sectors of the economy, transportation (56%), manufacturing (46%), and wholesale and retail (44%) are considered the sectors where workers are most at risk of losing their jobs. Overall, a study by Grace et al. (2018) at the University of Oxford paints a gloomy picture, suggesting that in about 120 years, automated machines will be capable of taking over all human jobs.

But AI Will Also Create Millions of Jobs

While the growth of AI will eliminate many jobs, it will create many more. However, many people find it incredulous to believe that AI will create new jobs. The fear is that AI will only take away jobs from people. They consider the present environment in which the COVID-19 pandemic has accelerated the automation of many routine tasks – which AI-powered robots currently do – and conclude that jobs would be lost and not created.

However, the current scenario is not different from the fears expressed just a few decades ago with the invention of the Internet. There were concerns that the emergence of the Internet would render millions of people unemployed and lead to the collapse of many businesses. Despite those fears, the Internet created millions of jobs and saw the emergence of new companies whose net worth ultimately trumped oil companies within a few short years (Bansal, 2021). However, it has to be pointed out that in terms of potential, AI can theoretically eliminate all blue-collar jobs, all white-collar jobs, and even all professional cum management jobs. This is only in terms of theoretical possibility. Nevertheless, many believe that this is inevitable in the long run. For this reason, many countries are today talking about government salaries for all their citizens to enable them to continue enjoying a living even when nearly all jobs are decimated in the future. In the short to medium run, AI will create more jobs than it could destroy (See **Figure 24** on the job landscape by 2025).

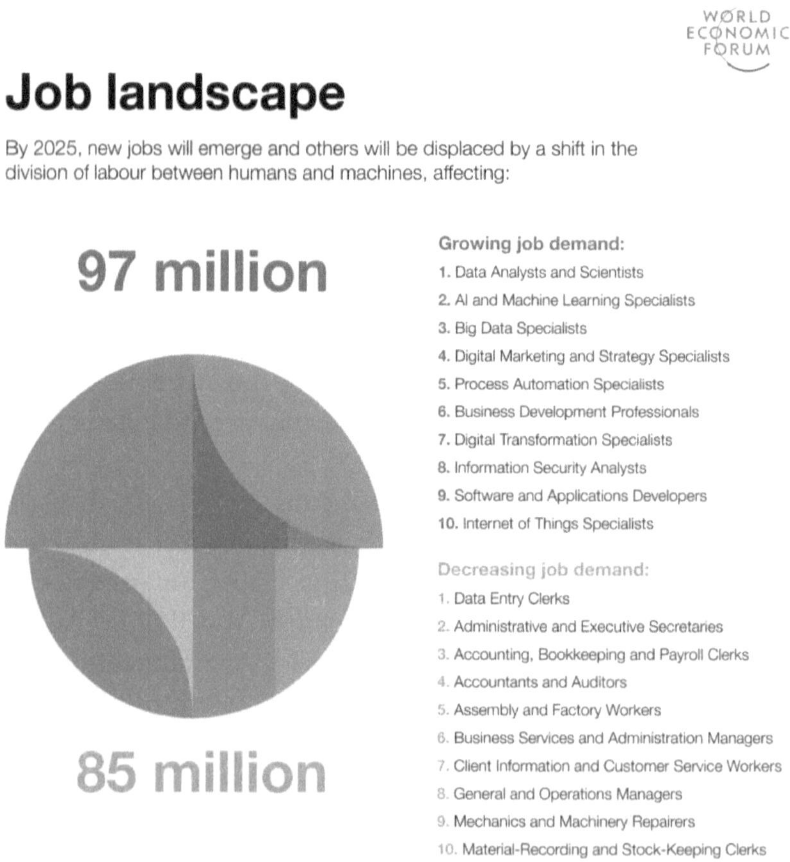

Figure 24: The Job Creation and Displacement Landscape by 2025
Source: Future of Jobs Report 2020, World Economic Forum

As the PwC (2017) research quoted earlier estimates, AI will increase global GDP by approximately $15.7 trillion by 2030 – a figure greater than China's and India's combined current GDP. According to the PwC study, while AI will automate some jobs in the process, "any job losses from automation are likely to be broadly offset in the long run by new jobs created as a result of the larger and wealthier economy made possible by these new technologies."

Aside from the study by PwC (2017), McKinsey Global Institute, in their research, opined that AI will displace approximately 400 to 800 million jobs by 2030. As high as those numbers are, the same report estimated that AI might create as many as 550 – 890 million jobs over the same period (McKinsey Global Institute, 2018).

Ultimately, it is becoming more apparent that AI growth will not lead to job losses but the emergence of more jobs and a higher global GDP. Chris Nicholson, CEO of Skymind, explained what the outcome of AI in work lives would be, thus:

> Everybody uses this analogy, but when the Industrial Revolution came, a certain kind of job disappeared.... However, many jobs, and many [new] jobs, were created. So when you think about England before and after the Industrial Revolution, it was not a poorer place where there was less work. There was a lot more work, but it was a different kind of work (Nicholson, n.d. as cited in Thomas, 2021).

The accelerated development of exponential technologies, of which AI is foremost, has triggered the fourth industrial revolution, and naturally, countries are striving to adapt to and harness these new technologies. The developed world is ahead of Africa in this race. As much as this is a massive challenge, it is also an opportunity. African countries could follow the example of the developed world, formulate new IT policies, make the needed investment, and eliminate unnecessary bottlenecks to spur technological development on the continent. That way, the continent will not be left on the fringes of the fourth industrial revolution, as was the case with the first three revolutions.

References

Accenture. (2016, September 28). *Artificial Intelligence Poised to Double Annual Economic Growth Rate in 12 Developed Economies and Boost Labor Productivity by up to 40 Percent*. Newsroom. Retrieved October 18, 2022, from https://newsroom.accenture.com/news/artificial-intelligence-poised-to-double-annual-economic-growth-rate-in-12-developed-economies-and-boost-labor-productivity-by-up-to-40-percent-by-2035-according-to-new-research-by-accenture.htm

Bansal, S. (2021, February 17*). Components of Artificial Intelligence - How it works?* Analytix Labs. Retrieved May 10, 2022, from https://www.analytixlabs.co.in/blog/components-of-artificial-intelligence/

Bughin, J., Seong, J., Manyika, J., Chui, M., & Joshi, R. (2018, September 4). *Modeling the global economic impact of AI*. McKinsey & Company. https://www.mckinsey.com/featured-insights/artificial-intelligence/notes-from-the-ai-frontier-modeling-the-impact-of-ai-on-the-world-economy

Emen, T. (2020, September 4). The potential economic effects of Artificial Intelligence. Turkish Policy Quarterly. Retrieved May 11, 2022, from http://turkishpolicy.com/article/1024/the-potential-economic-effects-of-artificial-intelligence_ftn16

Forbes Technology Council. (2019, March 1). *Tech experts predict 13 jobs that will be automated by 2030*. Forbes. https://www.forbes.com/sites/forbestechcouncil/2019/03/01/tech-experts-predict-13-jobs-that-will-be-automated-by-2030/

Grace, K., Salvatier, J., Dafoe, A., Zhang, B., & Evans, O. (2018). Viewpoint: when will AI exceed human performance? Evidence from AI experts. *Journal of Artificial Intelligence Research, 62(2)*, 729-754. https://doi.org/10.1613/jair.1.11222

Jewell, C. (2019). *Artificial Intelligence: the New Electricity*. Wipo. https://www.wipo.int/wipo_magazine/en/2019/03/article_0001.html

Lee, K.-F. (2018). *AI Superpowers: China, Silicon Valley, and the New World Order*. Houghton Mifflin.

PwC. (2017). *PwC's Global Artificial Intelligence Study: Sizing the prize*. PwC. Retrieved May 10, 2022, from https://www.pwc.com/gx/en/issues/data-and-analytics/publications/artificial-intelligence-study.html

Ronanki, R. (2021). *You and AI: A Citizen's Guide to AI, Blockchain, and Puzzling Together the Future of Healthcare* (p. 224). Forbes Books.

Singularity University. (2017, August 23). *Artificial Intelligence-Future of everything with Jason Silva (Part 6)*. Singularity Hub. https://singularityhub.com/2017/08/23/artificial-intelligence-future-of-everything-with-jason-silva-part-6/

Thomas, M. (2021, July 26). *Will a robot take your job? AI & The future of jobs.* Built In. https://builtin.com/artificial-intelligence/ai-replacing-jobs-creating-jobs

Chapter 7
Megatrends of the Future

From the time Charles Babbage invented the Difference Engine to the ENIAC, and ultimately the invention of the Internet, the World Wide Web, laptops, tablets, smartphones – and other hand-held devices – and presently, 5th generation computers, technology has continuously reshaped our world.

Presently, the exponential technologies discussed in chapter 5 are rapidly reshaping the world in a dizzying manner that is difficult to keep up with. Over the past several decades, a revolution in digital technology has completely changed every aspect of our personal lives and how business organisations operate.

In his book *Singularity Is Near*, Ray Kurzweil, futurist and co-founder of Singularity University, clearly portrays the quickening pace at which digital technology is progressing and has advanced an explanation for the exponential speed. According to his law of accelerating returns, the exponential rate at which technology has progressed is rooted in an evolutionary process akin to the type we see in biological organisms (Kurzweil, 2005).

In Biology, some members of each generation develop certain features – as a result of gene mutation and recombination – that confer survival advantages on them, which they pass on to the next generations. As this process plays out from one era to another, there is an incremental growth or improvement process. By building on this incremental progress, generation upon generation, organisms have advanced in complexity and capability over time instead of starting over.

In like manner, each generation of technology builds upon the innovative accomplishments of previous generations. According to Kurzweil (2005) and Russell (1992), this is a technological evolution in the world that has been occurring for millennia. New technology innovations do not reinvent the knowledge of previous technologies; instead, they build on top of their predecessors, thus, making for more rapid development of a new generation of technologies.

Technology Megatrends

We have seen many breakthroughs in technological innovations in the past several years. For example, we have seen tremendous new technologies like AI, blockchain, IoT, cloud computing, autonomous driving vehicles, AR/VR, 5G, deep learning, machine learning, 3D printing, and big data. The world is, thus, getting used to revolutionary new digital technologies that are quickly becoming part of our contemporary realities of life, at home, in our business, and in government.

While much innovation has already been achieved, current technology researches indicate that some interesting, emerging technologies will further redefine our future and change our lives unimaginably from just a few years ago.

Current trends indicate that multiple new technologies will emerge during the next two or three decades that will transform and improve human experiences and capabilities while creating disruptions across all strata of society. Given the magnitude of these changes impacting humanity, I chose to describe them as technology megatrends of the future.

Technology Megatrends of the Decade 2021 - 2030

The decade between 2021 and 2030 is poised to usher in many exciting technological advances. Some of the standard technologies have already started being used. Some others are in the making and will soon become part of our everyday-use systems within the next ten years. Some of the new exponential technologies are:

1. Artificial Intelligence (AI). Massive developments have continued in the AI domain in the past 20 years. However, in the years ahead, AI will continue to advance and remain integral to the operation of several other technologies. It is expected that global spending on cognitive and AI systems will reach over $97.9 billion in 2023 (Business Wire, 2019), while the AI market is projected to reach a total of $407 billion by 2027 at a compound annual growth rate of 36.2% (Markets and Markets, 2022). AI technology can assist in transforming Africa. They can be deployed in practically all sectors of the economies of all African countries. Moreover, if this is done urgently, it will significantly increase productivity, the number of goods in the respective economies, the living standards of people in Africa, improve the performance and productivity of public sector organisations, and enhance services in the education, health and hospitality sectors.

2. Internet of Things (IoT). The IoT refers to the ever-growing number of physical objects or things with Wi-Fi connectivity and inbuilt sensors that allow them to connect to the Internet and one another, thereby gathering and exchanging data over the Internet without human involvement.

We are already interacting with advertisements using the IoT. We can remotely lock and open the doors to our homes, turn on the oven in our kitchen to pre-heat it while we are on the way home or view the security camera along the fence of our homes or offices by simply tapping on a button on our mobile phone. For businesses, the ability of IoT to gather and analyse data allows for better decision-making and predictive maintenance.

Over the next decade, the impact of IoT on individuals and businesses will only increase. By 2030, the forecast suggests that the number of connected devices globally will be a 125 billion, while global spending on IoT will be $1.1 trillion (Vailshery, 2021).

3. Blockchain. With blockchain technology, we can now have more secure ways of securing our data, leading to more trusted transactions. This is because it provides a better approach for storing and authenticating data, assuring its protection. Blockchain data and transactions are immutable, as you can add but cannot remove or change them. It offers security that is very useful in facilitating trusted transactions. Over the next ten years, more and more businesses and government agencies will implement blockchain technology in their operations.

4. Cloud and edge computing. Cloud computing has been very useful in enabling businesses to store files and applications on remote servers and then access the data through the Internet whenever they want and from any location. Edge computing has emerged to take the game of data storage to a higher level.

It is expected that cloud computing will be so widespread by 2030 that it would be difficult to imagine that there was a time when it did not exist.

5. Wearables and Smart Clothes. Experts have projected that within the decade from 2021 to 2030, clothes and other wearables will be fitted with a chip that transmits various information about the wearer via the Internet. Thus, wearables will allow humans to perform better and to live with more health, safety and efficiency. It is forecast that by 2024, as much as 10% of people will own and wear smart clothes.

6. Autonomous or Self-Driving Vehicles. Currently, over 1500 self-driving cars are on the roads in the US (Kopestinsky, 2022). Carmakers in the US, including Tesla, are building vehicles with various fuelling options, including electric vehicles, hybrid cars, and hydrogen cars with self-driving technology, which is expected to be quite familiar by 2030. It is projected that self-driving cars will reach 20% – 30% of vehicles on the roads in the country. Between 2021 and 2030, the decade will see all kinds of autonomous vehicles – cars, taxis, trucks – on our streets. Autonomous vehicles will become commercially viable within the decade.

7. Space Tourism Can Become a Reality. The race to achieve space tourism and space travel is expected to gain momentum over the next decade. Some companies, such as Blue Origin, SpaceX, and Virgin Galactic, are developing suborbital tourist vehicles for conveying people to space. Already, there have been a few such travels to space. In 2021, billionaires Jeff Bezos and Richard Branson took suborbital flights to space in their respective supersonic rockets (Wattles, 2022). It is expected that the first tourist vehicles for civilian space tourism might be first tested by 2030, while official space travel will be introduced by 2050.

8. Big Data and Augmented Analytics. Big data refers to the large, difficult-to-manage volume of structured and unstructured data that businesses and government agencies generate daily in their operations. Augmented analytics is the process whereby vast amounts of data are gathered from original data sources and are thoroughly analysed, providing valuable insights concerning data that would have otherwise remained unanalysed without the help of analytics. So, with augmented analytics, organisations can now use enormously complex data from various sources. In the years to come, both big data and augmented analytics will become even more critical to the operations of businesses and government departments and agencies.

9. Intelligent Spaces and Smart Places. These are physical locations – such as offices, homes, factories, vehicles, and even entire cities – equipped with a network of sensors. Installing these sensors across facilities reduces operating costs significantly, increases productivity, and increases occupant comfort. Detailed information on occupant presence, location, count, and activity, in particular, may now be used to improve facility management substantially. The safety characteristics of the smart sensors include the capacity to help evacuate occupants in emergency scenarios and identify possible incidents. Tracking occurrences such as loitering and tailgating in pre-defined locations might increase security in areas where surveillance cameras are not present. In the next ten years, many smart spaces will surround humans in their homes, offices, factories, vehicles, and public spaces or buildings.

10. Digitally Extended Realities (XR). This is an umbrella term that describes immersive technologies – including virtual reality, augmented reality, and mixed reality. These technologies can merge the physical and virtual worlds to allow individuals to perform actions anywhere across the world as though they were there physically. The XR market is estimated to be about $209 billion in 2022 (Alsop, 2022). Meanwhile, in a survey by Raconteur (2019), more than 60% of respondents believed that XR would be mainstream in 2024.

11. Digital Twins. Many objects in the real world have their digital equivalents or representation. They are called digital twins. They use real-world data to create systems that imitate real-world realities. Therefore, specific, proposed changes or modifications that are desired to be made in the real-world object can first be trialled on the digital equivalent to determine the outcome on the real-world object.

12. Cyber Security. Cybersecurity might not necessarily be an emerging technology, but just like other technologies, it is evolving. As long as businesses and organisations face threats from hackers who seek access to their data, cybersecurity will remain a trending and changing technology. The ability to avoid cyber intrusions by hackers will be critical to the success of organisations over the next decade and beyond.

13. Quantum Computing. Currently, work on creating quantum computers is still primarily restricted to laboratories. However, it is expected that we will see the first set of commercially available quantum computers within the current decade. The worldwide market for quantum computing is predicted to be over $2.9 billion by 2029, while solid brands such as Microsoft, Honeywell, AWS, Google, Splunk, and others are investing heavily in quantum computing innovations.

14. Robots and Cobots. Many robots in use today are more intelligent than ever, thanks to the power of artificial intelligence. Without human assistance, robots will understand and react to their environments for a few years. As a matter of fact, in some industries, human beings and robots will collaborate as colleagues (Cobots) at work to get the job done.

15. 5G. The 5G networks will enable society to do much more quickly and easily. People will therefore browse the Internet, use social media platforms on their smartphones, use data-driven services, hold online meetings using applications like like Zoom, Google Meet, Microsoft Teams, and carry out streaming with the help of applications like YouTube and Spotify. It will come with faster, smarter, and more stable wireless networking, powering technologies such as IoT, smart cities, AR and VR, big data, and others. It will proliferate, and about 40% of the world will be covered with 5G in two years (Meek, 2018).

16. Virtual Reality and Augmented Reality. Virtual Reality (VR) enables users to be immersed in a digital reality as if they are in the digital space. In contrast, augmented reality enhances their environment by overlaying the real world with virtual information and objects such as images, text, and animation. VR and AR have been used in the entertainment industry for gaming purposes. In the decade between 2021 and 2030, these technologies are expected to become more integrated into our lives, especially in training, entertainment, education, marketing, and even rehabilitation, to aid recovery from injury. The global VR and AR devices market is expected to hit $209.2 billion by 2022 (Alsop, 2022).

17. Robotic Process Automation (RPA). This refers to the process of using robots to carry out some functions usually carried out by humans; for example, in repetitive processes like replying to emails, carrying files to the desks or offices of workers, repair of damages to a road, factory production, building constructions, interpreting health data, interpreting financial data and reports, and in the processing of varying transactions. In 2020, the global market for RPA was $2,078.3 million. This figure is expected to surge to $45,982.0 million by 2030 at a 36.3% CAGR (ReportLinker, 2021). Given that RPAs make the lives of different workers and process operators easier and more convivial, we can see why their deployment is gaining tremendous momentum every year.

18. Voice Interfaces and Chatbots. With chatbots, humans can have conversations (chats) with computers. This can be through written texts or spoken language. Some businesses have started adopting this mode of communication to interact with their customers, and it is envisaged that even more businesses will do so in the next decade.

19. Computer Vision and Facial Recognition. Computer vision is a branch of AI that allows computers to "see" by extracting information from images, videos, and other visual inputs. It uses it to interpret the world and make recommendations accordingly. Thus, facial recognition is used by computers to recognise or verify a person through their face. Biometric security includes facial recognition. Many businesses and institutions are already using computer vision and facial recognition. In the years ahead, the two technologies will become more ubiquitous.

20. Genomics and Gene Editing. With this technology, scientists can look at any organism's genome and edit it, i.e., add or delete some of the organism's genes. Various approaches are adopted for gene editing – one of which is CRISPR-Cas9. Through CRISPR-Cas9, the genetic make-up of organisms can be altered, which helps to correct DNA mutations that can, for example, lead to cancer. In

2019, the worldwide CRISPR gene editing market was put at $846.2 million. It is expected to reach $10,825.1 million by 2030 (Research Markets, 2021).

21. Machine Co-creativity and Augmented Design. This occurs when computers can carry out or assist in artistic works. Such creative works include works of art and music production. Augmented design is an aspect of machine co-creativity that utilises artificial intelligence and human intelligence to augment or enhance the design process. These technologies have vastly improved human creative processes.

22. Digital platforms. A digital platform is where producers and consumers congregate to exchange information, goods, or services and interact as a community. They include social media platforms like Twitter, Instagram, Facebook, TikTok, and LinkedIn. There are also knowledge platforms like Quora, ChatGPT, and Yahoo! Answers and service-oriented platforms like Uber and Airbnb. In the future, more platforms will emerge to disrupt more industries and turn established business models on their heads.

23. Drones and Unmanned Aerial Vehicles. Many organisations, even individuals, are allowed the use of drones. For example, military organisations use drones as aircraft piloted remotely or autonomously for their operations. They are also used for search and rescue missions, fire-fighting, law enforcement, and transporting materials into areas difficult to reach by other means. The usage of drones will only increase over time.

24. 3D and 4D Printing and Additive Manufacturing. 3D and 4D printing will have extensive applications over the next decade. It will be beneficial for mass personalisation.

25. Nanotechnology and Materials Science. This science is used to gain knowledge of objects at the atomic and sub-atomic levels. This way, scientists can create new materials and products at the nanoscale.

Technology Megatrends of the Decade 2031 - 2040

From 2031 to 2040, technological development will increase faster than what is obtainable presently. Such a massive transformation will offer us unique experiences and improve our capabilities as humans, enabling us to achieve a breakthrough in various areas of life. Some of the trends likely to prevail in the decade from 2031 to 2040 include the following:

1. Product Makers Would Embed Artificial Intelligence in Everything. By 2040, artificial intelligence will be embedded in virtually every device and permeate every aspect of life. Product makers would embed artificial intelligence in devices designed for healthcare delivery, transportation, education, logistics and supply chain management, hospitality services, security systems, and others.

2. AI Will Transform Industries and Labour Force. With artificial intelligence becoming mainstream, the technology will transform virtually every industry and disrupt the global workforce. It will eliminate many jobs while it will also create many job fields.

3. Data Will Be King. Artificial intelligence, the Internet of Things, and other technologies require large amounts of data to operate effectively. When artificial intelligence is embedded in everything, businesses in every sector will require large amounts of data to work efficiently and remain competitive. By 2040, an unprecedented amount of data will be available to offer valuable insight, which would be extremely valuable to businesses and organisations for decision-making.

4. There Will Be New Notions of Security and Privacy. Privacy ideas will continue to evolve as more personal information is required of individuals to access applications.

5. AI-Enhanced Weapons, Defence, and Security Systems. Over the next two decades, weapon makers will heavily incorporate artificial intelligence into existing weapons, defence, and security systems to enhance their performances. Likewise, countermeasures designed to nullify or confuse AI decision-making will likely emerge.

6. Smart Materials and Manufacturing Will Transform the Manufacturing Process. In the decade between 2031 and 2040, the production of consumer goods, building materials, and high-end military weapons will transform due to advances in material science and smart manufacturing. The transformation of the manufacturing process will allow new design options, reduce production costs, and extend product capabilities. The convergence of advances in artificial intelligence, nanomaterials, and quantum computing will accelerate the change facilitated by smart materials and manufacturing.

7. Biotechnology Enabling Rapid Innovation in Healthcare, Agriculture, and Manufacturing. The abundance of data, the ability to manipulate biological systems at the genetic level, and advances in automation and material sciences have opened the door to unprecedented innovation in the field of biotechnology – with applications in healthcare, agriculture, manufacturing, and cognitive sciences. With these innovations, there will likely be more abundant food for humanity in 2040, while many diseases that currently bedevil society will be eradicated or significantly reduced. Besides these, scientists expect the biotechnology sector to substantially contribute to the global economy during the next two decades, impacting as much as 20% of global economic activity.

8. Privacy and Anonymity May Become a Thing of the Past. It is likely that over the next two decades, privacy and anonymity may be tossed aside as government may track every aspect of individuals' private and professional lives, leading to a more open world. There will be a vast amount of information assembled on each person, so not much will be private anymore. It is likely that if some individuals choose not to make information about themselves available in the public domain, the government will mandate it.

9. Increased Cyber Vulnerability. On the one hand, the increased connectivity of individuals and organisations over the next two decades will most likely increase their vulnerability to cyberattacks. On the other hand, cybersecurity professionals would potentially develop new measures to counter cybercriminals looking to hack the hundreds of billions of connected devices.

Technology Megatrends of the Decade 2041 – 2050

Over the past several years, numerous technological innovations have wholly transformed many aspects of our lives. In the years ahead, there will be many more innovations. We cannot accurately predict the future. However, from the trends that we see in the present, we have been able to make very good inferences regarding the technological innovations that will likely prevail over the next decade and those that will probably be the mainstay of the decade from 2031 to 2040. From 2041 to 2050, computer scientists believe that artificial intelligence, the Internet of Things (IoT), blockchain technology, virtual reality, and other exponential technologies would have become the mainstay of daily life. Highlighted below are the possible technological transformations that will occur as a result:

1. Superhuman Clothing. There is the expectation that scientists will invent clothing that will give people super abilities. When worn, clothing will confer greater strength, allowing people to accomplish tasks far beyond their natural power. The exoskeleton that Hyundai, the South Korean company, recently launched is an excellent example of superhuman clothing. When completed, those clothing will be helpful for persons with health defects in their legs. With such clothing, even a paralysed man will be able to walk.

2. Generation of Renewable Energy Across Long Distances. Underwater cabling is improving daily, further enhancing Internet connectivity worldwide. Consequently, there is increased expectation that by this point in the future, countries will be able to generate solar energy from sources with intense sunlight – such as deserts – and convey the power generated through underwater cables to light up the entire country and other countries far away.

3. Hyperloop. This is a transportation system developed by Tesla. It is built underneath the ground level. It is a dome-like structure through which vehicles can travel through the tunnel using a high-speed propulsion system. Hyperloop is projected to become a global transportation chain in the decade between 2041 and 2050.

4. Space Vacations. Rocket companies such as SpaceX and Blue Origin are developing transportation systems that would make it possible for humans to take vacations into space. Elon Musk's dream of taking and bringing people back from Mars has already caught on with many people, and it is only a question of time before he achieves it.

5. Drone Solutions. Drones will become quite ubiquitous in the future. Scientists will develop drones with the capability to climb, fly and scan. These drones will enable people to go to places that humans cannot quickly go to today. Drones are already being used in Africa, especially Rwanda, to deliver products to people in remote areas, where it takes much more effort to reach.

References

Alsop, T. (2022, April 8). *VR/AR market size 2024*. Statista. Retrieved May 23, 2022, from http://www.statista.com/statistics/591181/global-augmented-virtual-reality-market-size/

Business Wire. (2019, September 4). *Worldwide Spending on Artificial Intelligence Systems Will Be Nearly $98 Billion in 2023, According to New IDC Spending Guide*. BusinessWire. https://www.businesswire.com/news/home/20190904005570/en/Worldwide-Spending-on-Artificial-Intelligence-Systems-Will-Be-Nearly-98-Billion-in-2023-According-to-New-IDC-Spending-Guide

Kopestinsky, A. (2022, March 5). *25 astonishing self-driving car statistics for 2022*. Policy Advice. Retrieved May 23, 2022, from https://policyadvice.net/insurance/insights/self-driving-car-statistics/

Kurzweil, R. (2005). *The Singularity Is Near*. New York: Viking Books.

MarketsandMarkets. (2022, August). *Artificial Intelligence Market Share, Size, Trends – [2022-2027]*. Markets and Markets. https://www.marketsandmarkets.com/Market-Reports/artificial-intelligence-market-74851580.html

Marr, B. (2021, July 2). *9 Technology Mega Trends That Will Change The World*. Bernard Marr. Retrieved May 11, 2022, from https://bernardmarr.com/9-technology-mega-trends-that-will-change-the-world/

Meek, A. (2018, November 27). *5G networks will serve 40% of the globe by 2024*. BGR. https://bgr.com/tech/5g-data-new-study-statistics/

Raconteur, (2019). *What is XR*? Raconteur. Retrieved May 23, 2022, from https://www.raconteur.net/infographics/what-is-xr/

ReportLinker. (2021, December 16). *The global robotic process automation market value is expected to surge from $2,078.3 million in 2020 to $45,982.0 million by 2030, at a 36.3% CAGR between 2020 and 2030*. GlobeNewswire News Room; ReportLinker. https://www.globenewswire.com/news-release/2021/12/16/2353303/0/en/The-global-robotic-process-automation-market-value-is-expected-to-surge-from-2-078-3-million-in-2020-to-45-982-0-million-by-2030-at-a-36-3-CAGR-between-2020-and-2030.html

Research Markets. (2021, February 8). *Outlook on the CRISPR gene editing global market to 2030*. GlobeNewswire News Room; Research and Markets. https://www.globenewswire.com/news-release/2021/02/08/2171285/28124/en/Outlook-on-the-CRISPR-Gene-Editing-Global-Market-to-2030-Analysis-and-Forecasts.html

Russell, P. (1992). *The White Hole in Time: Our Future Evolution and the Meaning of Now*. Harper.

Vailshery, L. S. (2021, January 14). *Internet of Things spending worldwide 2023*. Statista. Retrieved May 23, 2022, from https://www.statista.com/statistics/668996/worldwide-expenditures-for-the-Internet-of-things/

Wattles, J. (2022, January 3). *The year of space tourism*. CNN Business. Retrieved May 23, 2022, from https://lite.cnn.com/en/article/h_75a2e3d5132e1aed2c790f4efb402735

Chapter 8

Exponential Technologies and the Fourth Industrial Revolution

As established in chapter two, technology and innovation are the best drivers of economic development. The most unambiguous indication of this is the development strides made by the Western world over the past 150 years. These countries leveraged the technologies of the earlier revolutions to achieve their developed status. Unfortunately, Africa was left behind while the rest of the world moved ahead during those industrial revolutions. Consequently, Africa is quite behind the rest of the world in terms of both economic and technological innovation.

The fourth industrial revolution is unfolding now, underpinned by exponential technologies such as artificial intelligence, the blockchain, the Internet of Things, big data, 5G, and others. With the opportunities presented by exponential technologies, African countries can leapfrog their economic development to catch up with the rest of the world.

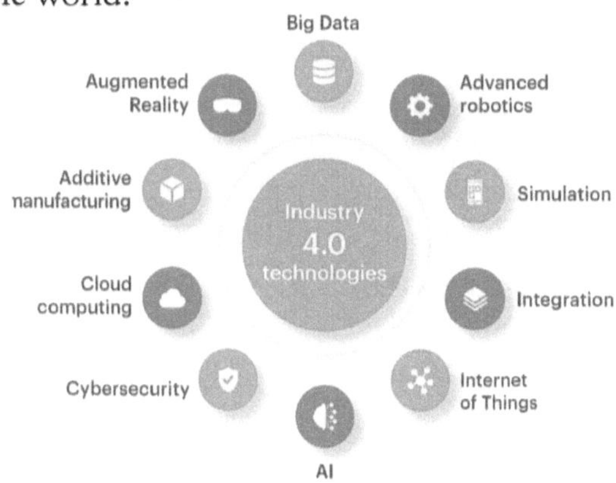

Figure 25: Fourth Industrial Technologies
Source: Business Process Incubator

Understanding the Fourth Industrial Revolution

What do we mean by the fourth industrial revolution? We must also understand the term "industrial revolution" to answer this question. We also have to examine the three industrial revolutions the world experienced in the past. "Industrial revolution" refers to a change in the production process, from handicrafts to factory machines, as in the first industrial revolution.

The starting point of the First Industrial Revolution (1IR) was in Great Britain in the middle of the 18 thcentury, from about 1760 to 1830. Many also agree that Wales was the first country to become industrialised. In other words, Wales industrialised first, meaning it was the first country to have a more significant number of its citizens working in the industrial sector of their economy rather than in agriculture. Gradually other European countries joined Great Britain in the First Industrial Revolution, followed by the United States of America, and some other regions of the world, including Canada, Australia, and New Zealand. The steam engine is singularly seen as the technology that catalysed the revolution of Great Britain. The revolution started as a result of improvements in the use of technology for industrial production, which led to large quantities of goods being produced in the country. With the steam engine in use from the late 1700s, the world moved from crafting goods by hand to mass-producing them in factories using machines. Industries such as flour mills, paper mills, cotton mills, ironworks, distilleries, waterworks, and canals could use steam engines leading to mass production. These changes brought about a massive boom in those industries.

Due to the economic boom, banks and industrial financiers rose to new prominence. Consequently, a stock exchange was established in London in the 1770s, while the New York Stock Exchange was founded a few years later, in the 1790s.

Further innovations led to another revolution known as the Second Industrial Revolution (2IR), which ran from 1860 to 1970.

During the Second Industrial Revolution, innovations in steel production, petroleum, and electricity led to the construction of rail lines which allowed cheap transportation of materials and products, resulting in an economic boom that, in turn, led to cheap rails to build more roads. The world could build solid steel structures like bridges, ships, aeroplanes, trains, weapons, cars, and skyscrapers.

However, the impact of the Second Industrial Revolution extended beyond innovation in steel production. During this time, apart from the production of steel, there were other notable inventions of the period: the harnessing of electricity, the telephone, the telegraph, the internal combustion engine, and automobiles.

As a result of the Second Industrial Revolution, there was a boom in productivity and economic activities in industrialised countries. Consequently, living conditions improved significantly. Also, the number of people engaged in agricultural activities dropped dramatically as more people who had previously lived in rural towns moved to urban centres to work in factories.

Further innovation introduced the Third Industrial Revolution (3IR) in the latter part of the 20th century, beginning around 1969. The invention of programmable computers and robots led to the automation of work and also to the automation of factory assembly lines. Apart from computers, the Third Industrial Revolution also brought us telecommunications and electronics.

The combined powers of the inventions of the Third Industrial Revolution turned the world into a global village. It changed the way people worked. It also empowered people to find, retrieve, share, and use data in ways that previously could not be done.

We are now in the era of the Fourth Industrial Revolution (4IR), which manifestly started in the year 2014. According to Schwab (2015), "It is a revolution that blurs the lines between the physical, digital and biological worlds while showcasing the advances in information and digital technology." These advances have provided a basis for a revolution powered by exponential technologies, including artificial intelligence, blockchain technology, cloud computing, the Internet of Things (IoT), robotics, big data, augmented reality, virtual reality, 3D printing, gene editing, and other technologies.

Following the arrival of the exponential technologies that are fuelling the 4IR, changes are occurring in a dizzying manner. Nothing is the same again, and yet we are to expect even more significant changes in our lives, how we live, at home, and at work, bringing about the disruption of practically every sector of society.

These new technologies provide African countries with the tools and opportunities to leapfrog development and catch up with the developed world as quickly as possible.

How the Exponential Technologies and the Fourth Industrial Revolution Can Transform Africa

In 2016, global business leaders and leading thinkers gathered at the World Economic Forum in Davos-Klosters, Switzerland, to brainstorm, exchange ideas, and answer the prevailing economic issues of the present era.

The annual meeting theme was "Mastering the Fourth Industrial Revolution." During the meeting, discussions on the Fourth Industrial Revolution took centre stage as participants shared new ideas for managing the pace, force, and scale of emerging exponential technologies shaping the world's economic, social, and cultural context.

There is no gainsaying the reality that the Fourth Industrial Revolution has arrived, fundamentally and exponentially changing the world as we know it. It holds the key to Africa's development. As a continent lagging behind the rest of the world in economic and technological development, the Fourth Industrial Revolution presents African countries with an opportunity to accelerate their growth. Some of the ways African countries can leverage the disruptive and exponential technologies of the Fourth Industrial Revolution to transform their economic fortune include the following:

1. Agriculture

Africa is blessed with abundant land that offers enormous potential for the continent's agricultural sector. In Sub-Saharan Africa, farming activities comprise about 60% of employment within the region (Goedde, Ooko-Ombaka & Pais, 2019). When the continent's youthful population is considered, the agricultural value chain is projected to add more jobs to the economy between 2020 and 2025 than any other sector. At the moment, the agricultural potentials of Africa are not fully tapped.

However, the Fourth Industrial Revolution offers an excellent opportunity to change the trajectory because it could provide farmers with information that would significantly boost agricultural production. Information on competitive pricing, prevention of crop diseases, mitigation of the impact of disasters, and the preservation of yields could lead to consistently high gains that would enhance farmers' income across the continent.

The agriculture sector is one key area that 4IR technologies are assisting in leapfrogging development. It greatly enhances the quality and quantity of agricultural products in the continent. For instance, in Ghana, agro companies such as Farmerline and AgroCenta provide farmers with farm advice, weather forecast, and financial tips via mobile-based web platforms, which farmers can easily access once they have signed up (Ekekwe, 2017).

In Nigeria, a startup called Zenvus is in the business of measuring and analysing soil data to help farmers know which fertiliser is most suitable for a particular type of soil. The information Zenvus provides also educates farmers on how best to irrigate their farms to ensure the highest yield. There are many other technologies across the whole of Africa today that are assisting farmers in having better access to market prices.

There are other possibilities that the Fourth Industrial Revolution can bring to the African agricultural sector. With the help of drones, AI systems, remote sensors, and blockchain, farmers cannot only maintain greater visibility of the situations on their farms but also maintain better control of things like water irrigation in their farms.

2. Energy

Energy production and distribution are important factors that will help Africa leapfrog economically. African countries suffering from energy deficiency can level up the electricity they generate using smart grids. With the help of IoTs and some of the plentiful renewable sources that abound on the continent, smart grids can be developed in various places, offices, cities and even homes, where they are consumed, rather than waiting for electricity to always come from centralised national sources.

3. Manufacturing

With additive manufacturing, as discussed in a previous chapter, products can be produced in proximity to where they are to be consumed, as explained above regarding smart grids and energy production.

4. Financial Growth

Following developments in new technologies like AI, blockchain and machine learning, banking and other forms of financial transactions are significantly transformed. We now have more innovative ways for carrying out banking online, as well as payments, online. This has enormously helped governments, financial organisations and society at large to take banking to many people at the bottom of the pyramid to have more inclusion in economic affairs.

It has thus created opportunities to provide financial services to people who could not be reached in the past. Also, these electronic payment platforms allow business owners to trade online and receive payments for their goods. Individuals who wish to can shop online and make payments online without the need to meet in person or visit a bank. In many countries, government agencies are also integrating online payment systems into their platforms to make it easier, faster and safer for citizens to make payments such as taxes and levies to the government.

5. Health

There are numerous challenges in providing good healthcare delivery in Africa due to various factors, including insufficient physical infrastructure and insufficiently qualified healthcare professionals. Nevertheless, the emergence of the technologies of the Fourth Industrial Revolution is poised to provide African countries with the route to successfully navigate these challenges and guarantee a sustainable healthcare delivery system for their citizens.

Already, mobile phone technology is successfully being used as a platform for gathering data to improve service delivery in some African countries. For instance, in Uganda, many public health workers use a mobile system called mTrac to track medication stock, mostly in rural communities. This ensures no shortages of malaria treatment and other essential medications in primary healthcare facilities. In Rwanda, the government uses drones to deliver transfusions to remote areas that are not readily accessible by road. During the outbreak of the COVID-19 pandemic, disease control centres across Africa used mobile phone services to disseminate information to the citizens about modes of infection, symptoms, diagnosis, and treatment.

In Ethiopia, artificial intelligence is employed in pharmaceutical production to help medical doctors accurately diagnose cervical cancer and various other ailments. IBM Research Africa is using artificial intelligence to discover the most effective ways to eradicate malaria from regions across the continent.

6. Smart Cities

IoT, AI and blockchain are three new technologies that are assisting the construction of smart cities in the world and, as such, would be immensely useful as Africa starts building smart cities. Smart cities are already being built in Kenya and Nigeria (Woherem & Odedra-Straub, 2017). Smart urban transport systems, one of the components of a smart city, lead to reduced traffic and road accidents and increases the ease of driving around a city. This will significantly reduce road congestion in cities like Lagos, Cairo, Nairobi and Johannesburg.

7. Education

There are numerous online courses today, some freely available on the Internet, powered by AI and other exponential technologies, providing rich and tailorable learning opportunities. Many platforms have pedagogical principles excellent for exploratory, dynamic, and creativity-based education. Those in Africa can make use of such systems to increase their access to education, especially as many do not have the funds to pay for expensive education from eminent universities and institutions abroad, and as some of those systems provide a virtual classroom environment that makes students feel they are in a real classroom learning from a teacher.

Overcoming Key Challenges Facing Africa

The Fourth Industrial Revolution presents vital opportunities for African countries to achieve more excellent development intentionally. Therefore, what is needed is for leaders of African countries to recognise the importance of the 4IR technologies, recognise the urgency for solving most of their problems, muster the necessary willpower needed to see projects through, and implement appropriate policies and strategies that would assist in propelling them in the use of these technologies. Below are some of the methods that African leaders should prioritise.

Developing a Continental Strategy for Harnessing the 4IR: Achieving the goals of the 4IR in Africa needs a continental strategy. Therefore, this book advocates the development of a comprehensive 4IR government master plan in each African country. This master plan will include policies and programmes making the continent 4IR-friendly. The master plan can only be achievable through the following measures:

Firstly, each African country should task its IT Ministries, Departments, Agencies and R&D institutes to work out 4IR frameworks that best suit the government. R&Ds of African countries should examine the 4IR readiness and policies to improve affordable access to ICT infrastructure and services. They must study broadband access policies/statistics, national ICT policies, competition, data and cyber security policies (Naito, 2018). The R&Ds should also examine demand-side readiness, digital literacy, affordability, etc.

Apart from tasking the R&Ds, African countries should ensure that their citizens are digitally ready; otherwise, their 4IR master plans will fail. Governments should ensure that their citizens, especially those who constitute the workforce, are digitally skilled and empowered. Governments should empower them to acquire skills that make them enjoy the massive employment opportunities in the digitally-connected world.

Thanks to a strategic partnership, Africa's national strategy on the 4IR harnessing will be possible. To make good progress, African countries should try as much as possible to rely on their effort to achieve their goals. However, they should not fail to use the assistance of international organisations where available. For example, African states should seek partnerships from international businesses and organisations, such as Microsoft Corporation, Google, USAID, AfDB, IMF, etc. The strategic partnership should provide Africa with exponential technology tools and skills to make the continent 4IR-compliant.

Fixing the Labour-skills Mismatch: The 4IR comes with some new digital skills, and it keeps ushering even more unique skills, thereby leading to the emerging-declining skills dilemma. Automation processes of the 4IR cause the issue of emerging talents and machines replacing many workers. So, where African governments strive to create jobs for the burgeoning youth population, they should develop technologies that enable young Africans to be employable in the 4IR digital labour market. Some existing technologies tend to replace low-skilled people—of whom Africa has an ample supply—with higher-skilled ones, limiting 4IR participation to economies with suitable capabilities. To guarantee that technology complements, rather than replaces labour, African countries must invest in education and re-skilling initiatives.

Agile Governance: No considerable success can be made in overcoming the challenges confronting the 4IR exponential technologies in Africa without agile governance. By enhancing elegant control, African countries can effectively manage the 4IR technologies to have a place in global value chains. To achieve these results, rule in Africa needs innovation. We know that innovation is at the heart of the 4IR; African governments need to reinforce state and institutional 4IR capacities to drive and support innovation and create the enabling socioeconomic environments for the exponential technologies of the 4IR to thrive. To exploit most of the benefits of 4IR tools towards enhanced development and closing the gap between Africa and the other continents, African leaders must ensure that they have the right policies and strategies that align well with those technologies. Also, as I have always maintained, governments can develop policies and procedures that are one of the best in the world; until they are religiously executed, the development will continue to be a mirage for Africa. The will to execute diligently must be there, which has not been the case in several African countries.

Developing Physical and Digital Infrastructure: Physical and digital infrastructural constraints in Africa, such as lack of energy, low teledensity, low Internet density, and low broadband adoption, limit access to advanced technologies. For this reason, the number of people with access to mobile phones and the Internet is still low in the continent. To overcome present impediments to Africa's technological growth, there have to be agreed standards for programming user interfaces, networks, programming languages to use, data formats, how data is collected and managed, how to reduce bias in data management, and AI for good development techniques.

Therefore, accelerating fibre-optic network physical connectivity and virtual platform interoperability is crucial for upgrading technology across the continent and reaching and cutting unit costs for the underserved. Appropriate infrastructure development will generally fuel and sustain Africa's economic transition. More African countries will be able to enter service export markets if the Internet connectivity and access gap between Africa and developed economies are bridged. Access to digital platforms for research, sales, and distribution may help African small-scale manufacturers become more competitive.

To make the most of the 4IR, African governments and entrepreneurs must identify new industrial niches and harness them to generate sustainable, inclusive growth, as well as take proactive efforts to close digital skills, infrastructure, and R&D gaps.

Overhauling the Educational System in Africa towards Exponential Technologies: The 4IR architectures need a digitally-demand-driven education system. Consequently, African countries' education systems and curricula need to be overhauled to meet the digital demands of the 4IR. The bitter truth is that majority of African graduates are unemployable digitally. In other words, their education has not equipped them with the modern skills of the 4IR. Therefore, they face unemployment in large numbers unless they earnestly and urgently start reskilling and upskilling themselves.

Overhauling the education curricula of African education systems should be a bottom-up exercise. In other words, the governments should start from primary school to universities. This is because learning is cumulative. This action puts enormous responsibilities on the government. They must assemble education planners, scientists, researchers, and ICT specialists. Those experts will map plans to ensure that the continent's education systems meet the 4IR standards.

Reshaping the curricula will need the services of teachers and lecturers. The African academics help develop local course contents that align with national curricula suitable for learning digital skills. It is inappropriate for most African learning institutions to rely on textbooks from other countries like the United Kingdom and the USA. As African education experts re-design the curricula, colleges and universities must form different schools and faculties that are basically about the 4IR courses.

References

Ekekwe, N. (2017, May 18). *How digital technology is changing farming in Africa*. Harvard Business Review. https://hbr.org/2017/05/how-digital-technology-is-changing-farming-in-africa

Goedde, L., Ooko-Ombaka, A., & Pais, G. (2019, February 15). *Winning in African agriculture*. McKinsey & Company. https://www.mckinsey.com/industries/agriculture/our-insights/winning-in-africas-agricultural-market

Naito, T. (2018). *Current state and development potential of the ICT sector in Africa*. New Breeze. https://www.ituaj.jp/wp-content/uploads/2018/01/nb30-1_web-05-SpecialAfrica.pdf

Schwab, K. (2015, December 12). *The Fourth Industrial Revolution: what it means and how to respond | World Economic Forum*. World Economic Forum. https://www.weforum.org/agenda/2016/01/the-fourth-industrial-revolution-what-it-means-and-how-to-respond/

Woherem, E., Odedra-Straub, M., (2017, May). The Potentials and Challenges of Developing Smart Cities in Africa. Circulations in Computer Science. Vol 2, No. 4 (27-39. https://wwwdoi.org/10.22632/cc-2017-252-06

Chapter 9

Most Potential Areas for Accelerated Adoption and Diffusion of IT in Africa

Information Technology is a disruptive technology that affects all areas of human endeavour. Its transformational impact on the economies of developed countries in the global north that have fully integrated it into their systems on an industrial scale is undeniably apparent. However, African countries in the global south have suffered varying degrees of underdevelopment in various sectors of their economies. Considering the role IT has played in developing the global north, this chapter examines the potential areas where IT can help the continent achieve accelerated economic growth.

Putting Technology in Africa into Context

The word diffusion is used in chemistry to refer to a process whereby particles, atoms, molecules, and water vapour move from more substantial areas of concentration to weaker areas of concentration. The terms "higher" and "lower" are significant in the definition of diffusion applied to technology in Africa. The term *higher* could be attributed to the developed economies where the widespread use of IT has resulted in impressive socioeconomic development. Meanwhile, *lower* is better used for less-developed or developing areas, where underdevelopment is a menace.

Regarding technology, we say that a particular technology has diffused into society when the technology is adopted and is being used in relevant areas in that society, even if the technology was initially developed in another part of the world. There is the need for the rapid adoption of technologies in various regions of the world, enabling private organisations and government agencies to take advantage of the latest technological innovations for executing tasks and promoting efficiency in operations.

Given the interdependent relationship between technology and development, it is imperative for less developed countries to urgently adopt new technologies from other parts of the world and, of course, to also develop new technologies locally through their local R&D and innovative efforts.

Generally, in the last 100 years, technological innovation occurred mainly in the world's most advanced countries, mainly in the West. Hence, technological innovations are diffused from developed to developing countries, mainly through foreign direct investment and the import of technology products by the latter.

Today, many hitherto developing countries are rapidly building up their local innovation capacities and rapidly evolving new technologies, thus industrialising rapidly. Examples of such countries are China, South Korea, India, Indonesia, Vietnam, and Singapore, in Asia; Brazil, Mexico, Argentina and Cuba in the Americas; and Turkey, Iran and Israel in the Middle East.

Therefore, the diffusion of technology in Africa means continuous and sustainable spread and application of IT in all sectors of the economies of all the countries on the continent, from "the least-developing countries" to "the faster-developing" ones.

The Impact of Diffusion of Technology on African Economies

IT is at the core of a "new" knowledge-based economy. It paves the way for more significant innovation, which drives productivity, growth, and the ability for countries to benefit from globalisation. All the above factors and outcomes are mainly visible in developed economies.

Africa's Information Technology sector snowballed in the recent decade, resulting in significant changes in the operations and processes of its companies. The IT sector has aided in the reduction of communication costs, the increase of market information, the formalisation of capital, the rise in productivity, and the facilitation of capital accumulation. This has resulted in lower costs of doing business in the continent. Though technology diffusion on the continent is not yet perfect, many indices indicate IT has been gradually building its presence in African economies, as can be seen in the following areas:

1. Digitalisation of Economies

Almost all sectors of African economies are becoming increasingly digitalised (Duarte, 2021).The continuous development of networks, sensors, artificial intelligence, and automation throughout Africa propels the revolution to a more developed destination.

Since the last 30 years, most of the indices of digitalisation of an economy have been increasing in Africa. In virtually all the countries of Africa, personal computers per head have increased in all the economic sectors. Similarly, smartphones, Internet connectivity, broadband coverage, teledensity, and use of social media have mildly improved in some African countries and in others.

The above is not gainsaying the fact that there is still a lot to be done. Compared to most other parts of the world, especially the more developed parts of the world, there are many grounds African countries still need to cover. When it comes to most of the new exponential technologies like AI, blockchain, cloud computing, robotics, IoT and quantum computing, African countries are either non-starters or are greatly lagging. Economies can be classified into five levels of digital maturity: Beginner, Basic, Operational, Strategic, and Exponential (see Table 1). In that case, most African countries are at the basic level of digital maturity. According to anecdotal evidence, some African countries, such as South Africa, Kenya, Rwanda, Nigeria, Egypt, Ghana, Morocco, Algeria, and Tunisia, are at the operational level of the digital maturity model. Most of the other countries in Africa are, at best, at the basic level of the digital maturity model (See Table 1). There are no countries in Africa at the exponential or even at the strategic level.

Table 1: Levels of Digital Maturity Model

Maturity Level	Name	Characteristics
5	**Exponential** Leverage exponential technologies to drive business growth. Business is disruptive in using Information Technology, emerging digital initiatives and business models. Adapt to the ever-changing digital trends and constantly leverages the same for business innovation.	• Satisfies all in level 4 • Must be an exponential organisation • Full-on cloud strategy implementation • Has a technology observatory • Employs exponential technologies like Artificial Intelligence (AI), Robotics, Machine Learning, Driverless Vehicles, Nanotechnology, Virtual & Augmented Realities, Drones, Quantum Computation, Cloud Computing and Bio-Engineering. • Big data and Analytics in routine use • Predictive analysis of future revenue • Constantly monitoring its processes and its ability to deliver new business models • Remakes existing markets and creates new ones
4.	**Strategic** Integrated, Synergistic, IT-driven organisation with its Business versus IT goals aligned to deliver digitally enabled services. Effective IT Strategy, policies and framework.	• Not only are operational systems computerised, but strategic systems needed for the competition are also in place • Automated process optimisation • Leans toward cloud implementation • Business Intelligence (BI) tools implemented • Use of metrics abound • Use of standard policies, principles and frameworks are routine • Driven by documented IT Strategy • Good data centre, data management strategy and data repository/warehouse. • World-class digital products, services and experiences

3.	**Operational** Technology plays a significant role in business operations. Core operational processes are digitised. Business and IT goals are aligned towards potentially disruptive technologies.	• Not yet identified which of their core function would benefit from digital transformation • IT Strategy in place • Core operations are computerised • Many other processes computerised are foundational or hygienic • Good data centre/network in place • Good security policy • Good business continuity processes
2.	**Basic** Business and IT initiatives are incoherent. Identified a need for a digitally enhanced, customer-driven business strategy, but tend to use it as the need arises (per project basis). Success is not predictable or repeatable.	• Some routine processes are computerised, e.g. Payroll, Word processing, Bookkeeping, Spreadsheet • More core operations are not computerised • Some digitally enabled customer experiences • Products are inconsistent and poorly integrated • Lacking defined company-wide standards and processes • Digital technology used to counter the threat
1.	**Beginner** Has no grasp of how digitisation would help drive business models and operations.	• Little or no use of technology • Nearly all processes are done manually • Weak customer experiences

2. Internet Penetration

There is now widespread usage of the Internet in Africa. This is perceived as a sign of Africa making progress to join the rest of the world in the Fourth Industrial Revolution and as a sign of increasing prosperity in the continent. Since its inception, there has been more than a fifty-fold increase in Internet use in the continent today, with more than a quarter of the people now using the technology (Ofure, 2021). In fact, by 2030, the ratio of people with Internet access in Africa will rise to more than a third of the continent's population. This will enable Africa to have a levelled playing field with the rest of the world in access to the Internet. This portends a high potential for the growth of Africa's economies following the high rate of Internet diffusion within the continent.

3. Mobile Technology

According to Allen (2021), the Internet has already generated 1.7 million new jobs and contributed about $144 billion to the GDP of Africa, which is about 8.5% of the GDP of the continent. Africa is the world's fastest-growing mobile phone market, with subscriptions rising from 16 million in 2000 to half a billion by 2010 (Aker & Mbiti, 2010). Mobile banking is one of the areas where mobile technology has a significant impact on the lives of Africans, especially the poor. Because of technological advancements, millions of underprivileged Africans now access financial services via mobile phones. Thanks to networks like Kenya's M-Pesa, Africa has jumped ahead of other regions to become a mobile, peer-to-peer finance centre. This is one area the continent seems to be leading the rest of the world, to the extent that about half of the world's mobile banking accounts are in Africa. Also, Sierra Leone, one of the world's poorest countries, has recently formed a Directorate of Science, Technology and Innovation (DSTI). This is a national financial data architecture with embedded automated financial instruments which aims to improve service delivery and minimise corruption.

4. The Dark Side of the Digital Economy

There are drawbacks to IT diffusion in Africa. Let us look at the issue of the affordability of the Internet. Indeed, without affordable Internet and stable power, many low-income Africans living in rural areas will be unable to access broadband Internet. A positive correlation exists between the level of access a family has to the Internet in Africa and the standard of living they enjoy. According to a study by Takaaki, Rogelio, & Carlos (2020), 3G Internet access was linked to a 14% rise in consumption and a 10% decrease in poverty in Senegal. As a result, countries that do not address Internet access issues risk limiting citizens' prospects, worsening already-significant inequality, and igniting regional, political, and ethnic divisions due to regional inequalities and poverty.

Some of the critical issues affecting the continent from reaping the benefits of exponential technologies in enhancing its economies are the dearth of a good number of people with digital skills, the low level of needed IT infrastructure, and the high technology costs. According to the International Telecommunication Union (2016), the Internet penetration rate was 87% in developed countries and 40% in developing ones, but unfortunately, only 25% in Africa. They observed that while telecommunications costs have fallen sharply in recent years, they remain higher in developing regions of the world; for instance, in Africa, where mobile phone and Internet costs are about four times the cost in developed economies.

The accompanying risks of digital technology are another factor that challenges the diffusion of digital technology in Africa. For instance, sophisticated IT enables different forms of criminality. With the help of IT, many African youths, especially those idle due to not being gainfully employed, indulge in all manner of cybercrimes. They are:

- **Online scams**: A scheme in which a person is persuaded to reveal personal or financial information in return for a fraudulent email or text message that falsely purports to be from a reliable source. The INTERPOL-supported Operation Sarraounia, which rescued 232 victims of human trafficking from Niger, including 46 minors, illustrates how social media facilitates migrant smuggling. As a result of the operation, 180 male victims were found to have been recruited online with messages that promised "decent employment" (INTERPOL, 2020).
- **Extortion using digital photographs**: victims are tricked into sharing sexually explicit images that are then exploited for blackmail.
- **Business email compromise**: Cybercriminals hack into email accounts to get information about company payment systems, then trick employees into transferring funds to their personal accounts.
- **Ransomware**: Cybercriminals block hospital and government computer systems and demand money to restore operations. Malware attacks are becoming more common in Africa. Cyber fraud costs increased by more than double in one East African nation between 2017 and 2018, totalling nearly USD 6.5 million (INTERPOL, 2020).

In addition, with the power of surveillance technology and

drones, new tactics of repression and the prospect of an autonomous weapons race loom.

Potential Areas of Accelerated Diffusion of IT in Africa

In the introduction of this chapter, I established the concept of diffusion of technology, stressing that diffusion of technology spreads from one area of high concentration to another area of lesser concentration. This diffusion does not leave any sector behind. Therefore, in this subsection, I will look at the potential sectors of the African economies where technology diffusion can be utilised. In examining these areas, I aim to highlight digital technologies in the selected regions, how they can facilitate operations in the mentioned sectors, create job opportunities for the teeming young population of Africa, increase the continent's GDP, and finally galvanise Africa to economic development, compared to the developed economies.

1. Agriculture

Most poor Africans live in rural areas and rely on subsistent farming for survival. Apart from this distressing fact, most African countries run a mono-economy, for instance, Nigeria, where the focus is oil. So, applying modern technologies in Africa's agricultural sector is paramount. The potential areas of agriculture where IT can help are:

The use of mobile technology in irrigation systems: In many parts of Africa, agriculture is hampered by inadequate rainfall and a lack of irrigation systems. Therefore, African governments can invest in using mobile technology to monitor and control crop irrigation systems. This technology can enable farmers to control the irrigation of their crops and farms with phones and computers. The technology could be simplified in the form of quick codes like USSD so even illiterate farmers on the continent can use it. For example, Nigeria has recently been suffering from perennial flooding caused by a combination of factors like climate change, the release of water from the Lagdo Dam in Cameroon, the overflowing of rivers Niger and Benue, and non-construction of needed dams, and poor irrigation in the various tributaries of the above rivers to needed farmlands. With the help of IT, flood conditions can be better monitored, controlled and used to manage the flood in such a manner that they can become yearly blessings rather than the curses that they are today.

IT & Agric financing: Most farmers on the continent are poor: they do not have money to embark on modernised agriculture. At times, governments provide grants and loans for farmers. However, in most cases, corruption makes it hard for farm grants, subsidies, and loans to reach the target farmers. IT can be a solution: African governments should invest in agri-fintech, where the poor farmers, through their mobile phones, can access finance for their farming (Joseph, 2021).

IT in farmers' security: Farmer-herder clashes are frequent in many parts of the continent. Apart from this, African farmers face different forms of attacks from thieves. With IT systems, farmers can better secure their farms and ensure their produce is safe. For instance, African governments can deploy drones to trouble zones.

IT & Livestock: IT can improve livestock farming in Africa. For example, with IT, DNA testing of livestock can be carried out, and meat conditions and level of quality can be verified using an ultra-scan. Also, big data can be gathered and used to make decisions concerning a farmer's livestock or aggregate figures for a country or the whole continent. Also, IT can boost the continent's agriculture by using gene-edited livestock that produces healthy offspring and meat.

IT & Crops: With IT, crop yield will dramatically improve. For example, IT can assist a farmer in checking the weather, determining when to plant, checking soil quality, monitoring the farm with the help of drones, deciding when to apply fertilisers, optimising nutrient absorption, and managing the farm daily. Similarly, IT can be used to modify crops to enhance crop quality and harvest genetically.

IT & Sale of farm products: The goal of commercial farmers is to make gains from their produce. IT can accelerate the sale and payment of farm produce in Africa. IT has the potential to connect African farmers with buyers of their products. Where the former does not need to go to the market to sell, he can use his phone, and buying could take place. For instance, 2KUZE is an IT tool used in Kenya, Uganda and Tanzania to connect farmers with agents, buyers, and even banks. Such technologies in widespread usage in all parts of Africa can significantly accelerate agriculture on the continent (Joseph, 2021).

2. Education

IT has revolutionised education in developed economies. There is what we call edtech or edutech, which is a combined use of IT and education to facilitate learning. African countries can use edtech to develop their educational systems in line with the digital world. This can be done in the following ways:

Increase in access to learning & teaching material through the Internet: One of the challenges confronting education in African countries is teachers' inability to obtain adequate instructional materials at all levels - primary, secondary and tertiary. The available materials are not updated.

Through the adoption of IT, such as the Internet, it will be possible for teachers, researchers and university lecturers to access the current information in the global networks. Thus, both teachers and students can access information beyond their local libraries but also from international sources.

Boosting online education: African countries seriously lack sufficient talented teaching personnel, funding to procure quality education materials, and adequate educational facilities and infrastructure. For instance, in many schools, the few available facilities cannot accommodate pupils and students; therefore, classes are crowded. Also, teachers are perpetually not enough.

These challenges can be tackled by investing in online education in Africa. With good broadband access to the Internet, many people in Africa can benefit from online education. They would be able to tap into the world's various intellectual and educational resources remotely. Thus, with online education, it will be possible to teach many African pupils and students who find it difficult to access education through traditional learning channels. Therefore, to boost online education on the continent, governments must increase education budgets to provide the relevant infrastructures to help online education thrive.

The use of localised pre-loaded digital devices: One of the best ways to apply IT to the education development of the continent is using tablets or devices containing localised and pre-loaded educational materials. The government of African countries can invest in such devices and make them available to African learners at all levels. The learning devices may be continuously connected to the Internet to update instructional materials.

3. Climate Change Adaptation

The application of IT in climate change in Africa can enable the continent to predict, analyse, and address issues and challenges regarding climate change. IT can help Africans achieve climate change adaptation in the following ways:

Access to climate change information: With the help of IT, Africans can have access to climate change information, thereby leading to capacity building in climate change issues. For instance, African governments can take advantage of the continent's rapid mobile penetration to ensure that Africans, even those in the remotest parts, can access data on climate change.

Monitoring the environment through IT: Climate change affects the quality of the environment, leading to water and food challenges. Investing in IT will make it easier for Africans to get information on weather forecasts, be updated about human-caused and natural climatic issues, and have sufficient data to provide early warnings on plant crops and potential dangers they face and needed remedial actions.

Carbon pricing systems: Carbon pricing systems also require the use of ICTs. Sensors, software, and networks can all be used to provide cost-effective market-driven solutions.

4. Healthcare

In Africa, relatively speaking, governments spend vast sums of their yearly budget on healthcare, but the expenditures mainly achieve nothing concrete. Therefore, health assurance is still very much of low quality in Africa. With the exponential technologies of today, Africa can revolutionise its healthcare infrastructure, making for more quality, efficiency and reliability. Below are some of the ways and techniques Africa can drastically transform its healthcare delivery system:

E-health: E-Health uses IT to support and deliver healthcare to patients and populations. Having seen how IT has transformed the health sector of the developed economies, it behoves African countries to overhaul their health systems using different e-health programmes. In other words, African countries can leverage IT to change their health systems from primarily analogue to digital. This means the application of computers, mobile phones, satellites, and software, among other things, in their healthcare delivery.

Mobile Health (m-health): Mobile connectivity is progressive on the continent. Most Africans, even those in rural areas, have mobile phones, even if the phones are not smartphones. So, African governments can leverage the availability of mobile phones to improve their health sectors. For instance, governments can design their health programmes using m-health so that citizens can access health services through their phones, irrespective of location.

Electronic Health Information Systems (HIS): Most health delivery systems in Africa are still manually processed via paperwork. Collecting, reporting, and exchanging data for quality and timely health data is challenging. The processes can consume time, energy, and money but result in inaccurate or incomplete data.

The use of IT tools can solve those healthcare delivery challenges. With an IT-based system, it will be easier to collate, analyse, and exchange data on patient records, infectious disease monitoring, manage drugs and supply chains, telemedicine communications with health workers, and many more. These measures will lead to efficient health data gathering and sharing.

Drone-Health (Dro-Health): Drones have been proven as effective suppliers of items to not-so-easily accessed locations. In Africa, drones can also be used in healthcare to supply medical equipment and drugs where there are inaccessibility challenges, and monitor supply chains, even to the remotest areas. Drone-Health technology can be most effective in times of emergency and outbreaks of infections. A country such as Rwanda is an excellent example of how drones can be effectively utilised to send drugs and other medical goods to very remote areas where they are needed.

5. Banking and Financial Services

IT makes it possible to "bank the unbanked." In other words, African countries can apply IT to enable Africans who do not have access to financial services to begin to do so. Therefore, IT can be used in the following areas of Africa's banking and financial services:

Data security: Banking and financial services in Africa are fraught with many data security issues, such as lack of trust, consumer protection, systemic network risks, etc. These issues slow the pace of the development of the sector. The good news is that IT can help to solve those challenges. It is left for Africa to invest in its population's financial data security needs, starting from individual, regional, national, and continental levels. With IT regulation, authorities can maintain a more water-tight control of financial systems. Financial institutions such as banks can digitise most of their operations and channels, or the end user touch points of their systems, leading to fail-safe access and use of their systems, with the use of high-end encryption techniques in data storage and communication of data.

Digital financial services (DFS) & financial technology (Fintech): DFS and fintech are some IT developments in the banking and finance sector. DFS comprises payments, remittances, and credits accessed and delivered through digital channels, such as mobile devices, the Internet, etc. On the other hand, fintech is a branch of IT that disrupts traditional financial methods in delivering financial services by infusing digital technologies into the banking and financial sector.

DFS and fintech can potentially transform Africa's banking and financial sector (Pazarbasioglu et al, 2020). These IT tools can be leveraged for cash-in/cash-out, bill payments, savings, remittances, insurance, and loans. There are many fintechs now operating in Africa, such as M-Pesa, Carbon, Opay, Flutterwave, and a lot more. Therefore, African governments can provide a conducive financial environment for more to thrive so ordinary Africans can be beneficiaries. Africa can achieve this by ensuring high-quality mobile phones and connectivity, well-functioning payment systems and interoperability, establishing credit infrastructures, expanding credit-relevant data coverage, supporting universal broadband connectivity, and aiming for high smartphone penetration.

6. Modernising Africa's Governance through IT

IT has facilitated electronic governance (e-governance). This involves the application of mobile devices, the Internet, and software for government efficiency. For example, in Africa, IT can be applied to ensure efficiency and integrity in the continent's electoral processes, as can be seen below:

Electoral data collation, storage, and retrieval: An election is one of the critical determinants of good governance. When citizens actively participate in electing their leaders, they can meaningfully demand good governance from them. Unfortunately, due to a combination of factors, for example, high level of corruption in the electoral system and the resultant voter apathy, voter turnout for elections have been very low in most African countries.

These issues of corruption in Africa's electoral systems can be addressed by investing in IT in the electoral systems. For instance, IT tools and applications can be deployed to capture, collate, store, transfer and retrieve electoral data in a highly secure manner. By so doing, the cases of massive rigging and the imposition by political leaders will dramatically reduce. So, IT can be applied to automate or computerise existing manual procedures to enhance access to and delivery of government services to citizens. If there is the will to do this, the technologies to use to make progress and to make the outcome tamperproof are available, for example, the use of blockchain, AI, and IoT.

M-Governance: The lack of inclusive governance is one of the continent's challenges. This challenge can be solved by m-governance, which involves using mobile technologies, such as SMS, USSD, mobile apps, AI, blockchain, etc., to provide accurate information and reach more trusted decisions, especially in the public sector. Mobile phones assist many more people with access to the Internet compared to the few currently accessing it through their computers.

Overall, using IT in the African government would improve governance, eliminate or reduce corruption, and assure the widespread participation of Africans in managing their countries.

7. The Application of IT in Africa's Regional Trade and Integration

One critical indicator of the economic might of developed economies is regional trade integration. This achievement is facilitated through the agency of IT. Africa can leverage IT in regional trade and integration in the following ways:

Provision of trade data: In Africa, regional trade and integration are hampered by a lack of relevant data. This problem can be sorted out through the use of IT, especially the exponential systems of today.

With IT, African entrepreneurs, private sectors, public sectors, and governments can invest in providing trade data for general consumption. The data will be able to provide information about who gets or who supplies what products and services –and at what cost, where, and the logistics involved. However, "scattered data" can be challenging to gather quickly; therefore, for Africa to truly achieve regional trade and integration, the continent needs unified databases.

As pointed out earlier, a database is an organised collection of structured information/data stored in a computer system and usually controlled by a database management system (DBMS). The combination of the data, the DBMS and the associated applications are called a database system or just a database. Africa needs many databases at organisational, regional, national and continental levels.

Unified database for regional trade and integration: As regional trade depends on cooperation between individuals, entrepreneurs, and governments of a community of countries, IT can help enhance regional business and integration. Just like Africa has regional and continental socioeconomic bodies, for instance, ECOWAS, ECA, AU, and others, establishing a unified regional trade and integration organisation can help leapfrog economic development on the continent. The AFCTA is a laudable initiative that aims to unify trade in Africa. It needs to be managed appropriately. Best initiatives, goals, and strategies are only good on paper unless they are brilliantly executed. They all must endeavour to ensure that the above good initiative is satisfactorily executed in order for the initiative cum goal not to be a mirage.

The use of blockchain in the supply chain: Blockchain is defined as a secure, trustworthy, and publicly available distributed ledger containing the details of every transaction and recorded proof of the proprietors of any value or asset at any given time. On the other hand, the supply chain is defined as the smooth and timely movement of goods from the producer to the consumer. So, blockchain is one of the IT tools that can be applied to ensure regional trade and integration in Africa.

The use of blockchain to manage trade supplies in Africa's regional trade and integration has numerous benefits. For instance, blockchain ensures accuracy, reduces data errors, guarantees efficiency and transparency, and has the potential to provide security and privacy while lowering risks and delays in the transit and delivery of goods and products. So, it assures trust in the whole process and outcomes.

The overall benefits: The application of IT in Africa's regional trade will provide revenue, employment, and investment opportunities and reduce Africa's economic dependence on developed economies.

8. Development of Smart Cities

A smart city uses IT to develop, deploy, and promote sustainable development practices to address growing urbanisation challenges. In other words, smart cities use Information Technologies to solve city challenges. The following areas below show ways in which IT can be used to take care of some of the difficulties African cities are coping with today:

Smart City IoT: Smart city IoT refers to the set of interconnected objects and devices with sensors that can automatically communicate and exchange data and information through a network. It includes electronics, firmware, software, sensors and various objects. All these devices will make it possible for African cities to share data on various points of interconnection. They can also provide needed intelligence and instrumentation through the agency of techs like wearable devices, smartphones, innovative electronics, and smart home gadgets.

Smart city analytics: This involves collating, manipulating, analysing, organising, and retrieving data for decision-making functions. The developed economies have embraced smart cities, and the massive socioeconomic benefits are enormous; smart city data analytics can turn raw data into alerts, signals, and information for the usage of Africans. A smart city can use data analytics to improve the monitoring and management of various city infrastructures while lowering risks and costs. Real-time data collected from sensors and other devices is used to assist in the optimisation of connections between modes of transportation for faster travel times, lower operating costs, and increase in user convenience through improved information services. Twende, derived from a Swahili word for "let's go," is a mobile app that utilises captured images from low cost-cameras and applies network-flow algorithms to estimate traffic flow. This means of decongestion does not require expensive road construction but the use of already present cameras coupled with deep analytics and specialised algorithms to interpret the data received (Woherem and Odedra, 2017).

Smart city governance: This happens through the use of the local and extensive data collected in the city to have insight into some of the city's critical issues and to make more informed decisions. It combines the previous two factors to create, administer, and regulate the operation of smart cities on the continent.

9. Manufacturing

Africa and other developing countries depend on more developed countries for manufactured goods. Fortunately, IT has the potential to change the continent's manufacturing dependence status. In other words, Africa can leverage IT in the manufacturing sector in the following ways: IT can be applied in the production of goods and products. These particular areas of application are discussed below:

The use of CAD/CAM: CAD means computer-aided design, while CAM means computer-aided manufacturing/modelling/machining. With CAD and CAM, computer systems can be used to design and manage products, from prototypes to finished product design, to manufacture, and to manage the manufacturing process operation for developing the product. In this era of 4IR, CAD/CAM is driving manufacturing, primarily through 3D printing and modelling. Currently, clothing and textiles, mining and exploration, building and construction, engineering, and other manufacturing industries are taking advantage of CAD/CAM technologies. Therefore, this aspect of IT can help Africa leapfrog economic development.

Computer-integrated manufacturing (CIM): It is a combination of IT and factory automation. It combines development in the manufacturing industry with IT to achieve a competitive advantage. It assists companies in automating the design, manufacturing, planning, control, and production of their products.

Automation and Control Technologies: The 4IR comes with automation and other control technologies. To be part of the 4IR, manufacturing industries are turning to automation. The manufacturing sectors in Africa can also use automation and control technologies. Through automation, powerful processors, more intelligent, lower-cost sensors, and robots are used for turning raw materials into finished products.

10. Transportation

Africa's transportation systems can leverage IT in the following ways:

Intelligent transportation system (ITS): "ITS" refers to the use of IT in transport. With ITS, the transportation system or network of a city or country can be seamlessly managed in such a manner that, for example, road and rail transport users are provided with up-to-the-minute information concerning transport schedules, the condition of different roads, rails, and trains, traffic situations, stoppages, which buses or trains are running, which are being delayed, travel durations to other places, etc. Africa can leverage this field of IT to improve its transport systems. This is a component of many smart city initiatives and can now be seen in many countries that have smart cities. For example, in Barcelona, South Korea, UK and Republic of Ireland.

Augmented reality (AR) and virtual reality (VR): Many transport industries in developed countries are experimenting with AR and VR in their transport sectors for picking passengers, luggage, inventory management, training staff, etc. Though VR & AR in transport are still in development stage, the two IT technologies are potential areas of application in Africa's transport systems.

Smart Transportation: This refers to the application of modern integrated ITS (intelligent transportation system) in transportation systems. The smart transportation system of a country encompasses the use of big data and AI to integrate and manage all the stakeholders, objects, users, and other actors in the network, including vehicles, traffic signal control, automatic recognition of vehicle number plates, and the use of speed cameras. Therefore, smart transportation has solutions to most of Africa's transportation challenges. This is, once again, one of the key aspects being implemented in various smart cities in the world.

11. Privacy and Security

Whereas Internet connectivity has many benefits for Africans, there are also downsides. One of the downsides is the breach of personal data and security. Websites, apps, and other login platforms will require permission to collect personal data to improve their operation. However, some of them violate users' privacy and security. For instance, some loan apps in Nigeria require access to users' phones to check whether the users qualify for loans. Unfortunately, some of them collate their contacts so that in the event of default, they will send messages to or call most of the people in the contact list. So, IT can pose a threat to Africans' privacy and security. On the other hand, IT also has the potential to safeguard them from privacy and security breaches. The following can be done:

The use of the Computer Response Team (CERT): CERT is one of the IT tools that can combat data privacy and security breaches. It is applied for detecting, investigating, and monitoring cyber incidents. It can also coordinate government recovery from cyber-attacks, ensure cybercrime preventive measures, and so on.

Data security: Data security refers to the actions, tools and processes utilised to protect data from unauthorised access and theft by individuals or systems that do not have the proper access rights to any form of data, be it data in databases or data sent across a network from one device to another; as well as the protection of data from users with malicious intentions and from organisations that own user data and may try to use them for ulterior intentions. Data security also involves protection of data from corruption. To assure data security, hardware, software applications, databases, storage devices, networks, etc, must be protected. One of the safest IT tools or techniques for African governments to assure data privacy and security for their citizens is through data security.

12. Building & Construction

IT has transformed building and construction in developed economies. Therefore, there are potential areas of IT applications in building and construction for Africa; they are as follows:

Drone & robots: Drone and robots are some 4IR technologies that African countries can embrace in their construction industries. For instance, drones can be used to conduct site survey quickly than a crew on the ground. They can use high-resolution cameras and collected data to create interactive 3D or topographic maps or models. As for robots, there are construction robots for brick-laying, masonry, and robots that lay an entire street at one time.

AR & VR: Though they are still emerging techs in building technologies, they have the potential for leapfrogging building and construction in Africa. For instance, AR can be used in safety training, real-time building project information, and many more. On the other hand, VR can be used to understand and evaluate complex building projects.

3D printing: This is a potential area where Africa can leverage to leapfrog development in the building and construction sector. It would change how materials are sourced and used in production or manufacturing. For instance, with 3D printing technologies, building and construction materials can be printed and transported to job sites, ready for use immediately.

13. Mass Media

The media industry in Africa can benefit from the application of IT. Below are some of the ITs that can help the continent leapfrog its media development:

Artificial intelligence: AI has permeated all fields of human endeavours, including the mass media. Multinational mass media companies in developed economies understand the impact of AI and they are aggressively leveraging it. For instance, Warner Bros has started using AI technology to manage its films and budgets.

In the same vein, the mass media industry in Africa can leverage AI. They can use it for fake news filtering, as the Internet is full of fake news, making it hard for readers to separate fact from fiction. Also, with AI, efficiency can be achieved by media practitioners.

Automation: News automation is one of the latest IT trends; it reduces costs and increases viewership. With automation, journalists can do their research and analyse data more efficiently and write and distribute media more easily through what can be called "automated journalism." Journalists now use AI algorithms to produce stories at scale. The African mass media industry can also apply this IT tool.

Blockchain technology: Blockchain technology presents vast potential for the mass media in Africa. For instance, the application of blockchain can bring about the independence of the mass media, where it is hard for non-partners to interfere in their data. This is because blockchain is a decentralised platform; it allows users to own their identity and data. This ownership can level the playing field between media companies and Internet giants who have been controlling user data and identities.

14. Security

Many African countries are bedevilled by various security challenges today such as banditry, kidnapping, cyber-crime, ritual killing, suicide bombing, and terrorism. IT has the potential to help the continent combat most of the above security risks through the agency of the following technologies:

Drone: Drones are one of the safest technologies for defence surveillance and for the supply of light military hardware, among other uses. Investing in drone technologies for defence in Africa has many benefits, chief of which are its stealth mode and the fact that it reduces costs and need for human boots on the ground.

Blockchain: African countries usually complain about mismanagement and corruption in their national defence, as military budgets are usually misappropriated. The use of blockchain in Africa's security architecture can reduce crime to minuscule levels, assuring trust, which will aid correct decisions to fight insecurities.

3D printing: Like in other areas, African countries depend on developed economies for their military supplies: software, hardware, equipment, and kits. Due to its ability to be used for additive manufacturing, 3D printing can change the dynamics of producing many military hardware and technologies, as a lot of what is needed to be procured from abroad can be built locally through additive manufacturing, thus reducing the cost of procuring defence products. This is a new technology. It cannot yet produce exclusive complex products, like tanks and missiles. However, it is evolving and would be helpful when fully developed.

Determinants of Technology Diffusion in Africa

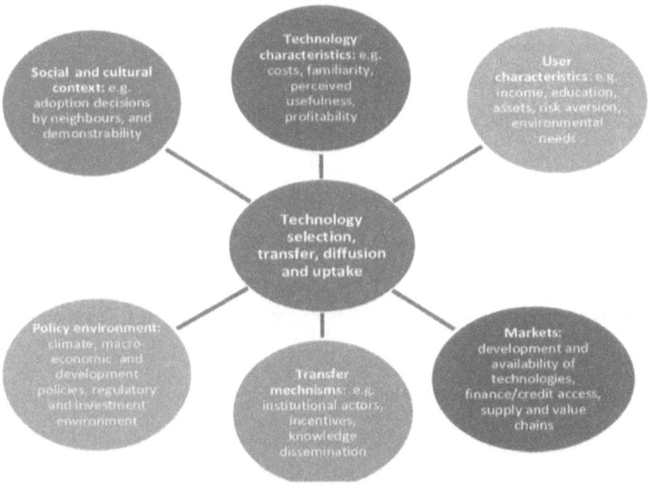

Figure 26: Technologies for adaptation and their transfer, diffusion and uptake, adapted from Anne Olhoff, https://doi.org/10.13140/RG.2.2.24250.80320

From Figure 26, we can see that some factors constrain the diffusion of technologies. Therefore, technology diffusion in Africa depends on numerous factors; it does not occur in a vacuum. But for the context of this chapter, we subsume the factors above into three areas as:

- Continental/Governmental IT Efforts

Government/governance is a macrocosmic powerhouse that drives other sub-power houses. Therefore, technology diffusion is impossible in Africa without concerted and sustainable continental/governmental efforts. The effort comprises IT policies, IT-R&D, provision of IT infrastructures, both physical and human, and more. It also includes the will to digitalise the economy, human capital availability in terms of numbers and quality, financial resources availability, and technology practice centres. For the technologies to be readily diffused into African societies there has to be intention on the part of the government, and even private sector organisations, to bring them in. It requires a national IT policy. It requires a centre, department and ministry responsible for driving IT procurement, development and diffusion in society. It requires effort to identify appropriate technologies and customise them to be suitable for the local environment. However, the government's actions are not enough without citizens' support.

- Population Characteristics

It is one thing for governments to make provision of factors that necessitate technology diffusion, and it is another for the citizens to embrace technology. For sustainable technology diffusion, the population within a given terrain must be willing to seek or embrace IT knowledge essential for technology application and development. And in all of this, there has to be motivation and incentive for the citizens to invest their time in IT.

- Technological Innovation

Every continent, country and place has peculiar technological needs. Applying the IT templates of the developed economies originally without remodelling is not practical for Africa. There should be home-grown innovations that will modify IT tools and applications to suit the socioeconomic needs of the continent. Through the importation of technologies from abroad and home-grown technological innovation, developing countries in Africa can not only start making good use of technologies in their operations but can also develop innovative technologies locally. This will lead to bridging the technology gap between Africa and other developing economies and the rest of the world.

Guidelines for IT Diffusion in Africa: Establishing the Role of Governments in Facilitating the Diffusion of Digital Technology for Development in Africa

In summary, this sub-section entails two things: technology diffusion does not happen by chance. Secondly, it happens by design, and not just by individuals, but by governments and, by extension, a continent. Therefore, in this chapter, I propose a four-part guideline for technology diffusion on the continent: technology design, technology implementation, adoption/adaptation, and technology diffusion.

Technology Design: The first step African governments can take in their efforts to ensure technology diffusion on the continent is to design their national IT architectures, policies, and programmes. Specific IT needs of the countries will guide these policies. Apart from technology design at national levels, Africa should come together at regional and continental levels to design IT frameworks which will result in the desired economic development. It is necessary to stress that developing continental IT policies and programmes for Africa involves extensive IT R&D.

Technology Implementation: Technology design is just a paper; it needs to be implemented. Technology implementation in Africa comprises such actions on the part of the governments: enough IT budgetary allocations, IT regulations that are flexible and accommodative of entrepreneurship and investments, training and re-training of IT personnel, provision of IT infrastructures, vast broadband availability, fibre optic, 5G network, stable electricity, high-speed Internet connectivity, and affordable data, among others.

Technology Adoption: Technology adoption here refers to when technology is selected for use by individuals, organisations, countries, and so on. It can be seen as a systematic and successful integration of IT in various sectors of Africa's economies. Technology adoption on the continent does not occur at a single stage; IT adoption in Africa will follow the usual worldwide steps of adoption of technology:

i. **Innovators:** They are eager to try and adopt new technologies. They are willing to take risks.

ii. **Early adopters:** They closely watch for innovations in the market, but they are notably more selective than innovators when it comes to making purchasing decisions. The innovators and the early adopters gain most of the benefits of the new technology as they are the earliest adopters. Of course, they also bear most of the negative impact associated with the technology.

iii. **Early majority:** They are more conservative and risk-averse in their decisions to adopt new technologies. But they eventually adopt new technologies due to its inevitability, alongside the majority.

iv. **The late majority:** Due to a high level of scepticism about the benefits of new technologies and having less financial flexibility than early adopters, the late majority adopts technologies much later than others.

v. **The laggards:** They are averse to new technological changes and unaffected by opinion leaders. They are therefore often the last to adopt new technologies.

Technology diffusion: The success of the three stages of technology design, implementation, and technology adoption and adaptation will culminate in technology diffusion in Africa. The term refers to the stage in which the technology spreads to general use and application in all sectors of Africa's economies. Currently, Africa is a laggard in most aspects of technology adoption, while it also is a pioneer (an innovator or early adopter) in some other areas, notably in mobile banking and finance, e.g., the M-Pesa in Kenya.

References

Aker, J. C., & Mbiti, I. M. (2010). Mobile phones and economic development in Africa. *Journal of Economic Perspectives*, 24 (3): 207-32. Retrieved May 11, 2022, from https://www.aeaweb.org/articles?id=10.1257/jep.24.3.207

Allen, N. (2021, March 11). *The promises and perils of Africa's digital revolution.* Brookings. https://www.brookings.edu/techstream/the-promises-and-perils-of-africas-digital-revolution/

Duarte, C. (2021). *Africa goes digital*. International Monetary Fund. Retrieved May 11, 2022, from https://www.imf.org/external/pubs/ft/fandd/2021/03/africas-digital-future-after-COVID19-duarte.htm

International Telecommunication Union (ITU). (2016). *ICT facts figures 2016.* International Telecommunication Union. https://www.itu.int/en/ITU-D/Statistics/Documents/facts/ICTFactsFigures2016.pdf

INTERPOL. (2020, August 14). *Online crime in Africa a bigger threat than ever before, INTERPOL report warns*. INTERPOL. https://www.interpol.int/en/News-and-Events/News/2020/Online-crime-in-Africa-a-bigger-threat-than-ever-before-INTERPOL-report-warns

Joseph, A. (2021, December 8). *Top 6 Innovations in Technology That Can Improve Effectiveness of Agriculture Industry*. Krishi Jagran. https://krishijagran.com/blog/top-6-innovations-in-technology-that-can-improve-the-effectiveness-of-agriculture-industry/

Ofure, E. (2021, October 18). *Tech Opportunities for Africans*. Techibytes. https://techibytes.com/tech-opportunities-for-africans/

Pazarbasioglu, C., Mora, A. G., Uttamchandani, M., Natarajan, H., Feyen, E., & Saal, M. (2020). *Digital financial services*. World Bank. https://pubdocs.

worldbank.org/en/230281588169110691/Digital-Financial-Services.pdf

Takaaki, M, Rogelio, G. O. & Carlos, R. C., (2020). *Broadband internet and household welfare in Senegal*. Policy Research Working Paper; No. 9386. World Bank, Washington, DC. World Bank. https://openknowledge.worldbank.org/handle/10986/34472

Woherem, E., & Odedra-Straub, (2017, May). The Potentials and Challenges of Developing Smart Cities in Africa. Circulations in Computer Science. Vol 2, No. 4 (27-39. https://wwwdoi.org/10.22632/cc-2017-252-06

Chapter 10

Socio-Cultural Implications of IT in Africa

Chapter 5 highlighted the fact that technology has existed for a long time and has facilitated the development of humans through the ages. Technology has evolved from one form to another, starting with the pre-mechanical age, the mechanical age, the electromechanical age, and now the electronic age. Alongside this evolution, several inventions were made, laying a foundation for further development. The emergence of the technologies that ushered in telecommunication, the design of the ENIAC, and subsequent technologies such as the Internet and the World Wide Web led to the emergence of the field of information technology.

Since information technology emerged as a distinct field, technology has further evolved, giving rise to numerous inventions. Today, exponential technologies represent the highest level of technological advancement. These technologies sparked the Fourth Industrial Revolution and significantly changed how people interacted, lived, and worked.

The world is undergoing a new industrial and technological revolution due to the impact of exponential technologies, resulting in the massive, quick, and broad transformation of society and industry. This shift causes both the manufacturing process and the number of products and services produced to accelerate. Information and knowledge are replacing labour, raw materials, energy, and land as the primary factors of production. New goods and services are being promoted, and the nature and structure of work are changing.

The current technological transition is blurring previously established divisions between manufacturing and service-sector occupations, between work, education, and leisure activities, and between male and female job responsibilities. In fact, exponential technologies such as artificial intelligence, the Internet of Things, blockchain technology, big data, robotics and others are pervading virtually all forms of human endeavour: work, education, leisure, communication, production, distribution and marketing. It also affects the extent and content of information networks, the interconnectedness of companies, and how individuals live, work, shop, study, communicate, and play.

The computer today is providing developed countries with a new, all-purpose metaphor or paradigm for perceiving humans and their relationships with one another and the world in general. The information processing model now dominates cognitive psychology. Humans are perceived as merely processing information with their "wet brain-ware" just as computers do with their hardware. Computers will soon be imbued with intentionality as humans do and have all human cognitive attributes. The human mentality will soon be seen with an information-processing metaphor.

Many view these future technological scenarios with much trepidation and horror, even in the West. Science has unwittingly removed God from people's lives, as it is perceived to challenge, through concrete empirical outputs, all that religion claims, or the foundations upon which the marvels of the world stand and wherein belief in the explanations of religion stands. Through science —for example, the Theory of Evolution and discoveries in physics, biology and archaeology —former beliefs are eroded, making people start questioning or losing interest in the explanations of religion.

To some philosophers of science, science has left the world, humanity and existence bereft of meaning and tries to install itself as the ultimate meaning or the religion wherein people can learn of the nature of things (Russel, 2004; Meyers, 2021). "Meaning" here refers to the most treasured primordial, instinctive understandings of the world that every society has developed through their unique cultural and spiritual spectacles, and which keep them in close contact with Gaia (the world) and their natural surroundings. In other words, those meanings give purpose to their lives. Science has thus arrogated to itself the right to provide explanations for everything and debars people's inclination to seek supernatural beings for explanations or theory, casting negative doubts as to the existence of any such supernatural being. It claims that the universe happened by chance, as Darwin and his most recent disciple, Richard Dawkins, propagate. Science now purports not only to tell humans how the world works but also why, without any "God" in the picture.

Many have summarised the paradigm-shifting onslaught of science thus: through Galileo and Copernicus, humans ceased to be in the centre of the universe; with Darwin, humans lost their unique position on earth; and with Freud, humans even lost control of their minds. So, perhaps, science only illuminates some aspects of our reality; it cannot understand many other things with its reductionist methodology.

Reductionism is a crucial scientific methodology. From a reductionist perspective, science examines the world and the phenomena it studies by reducing the whole to its constituents, then to the elements of the components, and so on, until it reaches irreducible particles or components. In physics and chemistry, for example, even atoms are reduced to electrons, protons, and neutrons, and eventually to elementary quantum particles like photons. When you limit what you study to only those things you can mathematically grasp, some of the other equally or even more significant properties may be overlooked.

Schumacher (1977), in his book *A Guide for the Perplexed,* argues that the world has inanimate objects, plants, animals, and humans as part of its order of existence. Inanimate objects are lifeless and, therefore, ideal for scientific investigation. It is possible to break down whatever you would like to study into its component elements and utilise scientific principles to examine what it contains, how it works, how it interacts, and what it yields. Science has excelled at inanimate objects. With it, science discovers and comprehends a variety of natural phenomena like galaxies, gravity, black holes, dark matter and energy, stars, and planets. It also helps scientists understand why celestial bodies are held up and appear not to be falling, as well as the weather systems and parts of various objects that are used to build bridges, create the computer and the Internet, map the minerals found in other parts of the world, and determine when the universe first came into existence. In other words, science has given us enormous power to exploit our environment and increase the quantity and quality of goods we consume as humans, which many see as a net positive for humanity.

When we look at plants, we see that they have something that inanimate objects do not have: life. As a result, a plant can die and become inanimate. Moving on to Schumacher's next key component of existence, we have animals. In addition to life and inanimate parts, animals have consciousness. Thus, an animal can be knocked unconscious and still be alive. If it dies, it loses its energy and becomes an inanimate cluster of organs, tissues, and cells. At the top of Schumacher's hierarchy of the environment on earth sits humans. Humans have metacognition (or self-consciousness). As a result, humans have life, consciousness, and self-consciousness. In other words, humans can be conscious of their consciousness. They can lose their self-consciousness while remaining conscious. They can still live even after losing consciousness. If they die, they become inanimate - in other words, a cluster of organs, tissues, and cells, without life.

Thus, as was said earlier, science is mainly suitable for studying the inanimate components of existence. It has yet to be effective or, in fact, largely ignores the study of life, consciousness and self-consciousness. This means that science is ineffective today in understanding what makes plants alive, animals alive and conscious, and humans conscious and self-conscious. However, science is used by many as their authority to remove from the picture all the seeming nebulous or qualitative things like life and consciousness. For this reason, it is maintained above that science is good at studying and understanding certain things due to the power of its reductionist methodology but is seemingly unable to research and adequately understand many other things. Science is thus like somebody who uses it (science) as a torch to illuminate some parts of their paths at night and uses what he sees to generalise the whole universe. So, yes, whilst I can see and believe in the enormous power of science and technology, I can also see that science does not illuminate everything that needs explanation in the world, and therefore, that we should not use only the descriptions of science to generalise or derive explanations for everything.

Many believe that the modern scientific and materialistic society, despite its economic, technological and military triumphs, shall lose or is already losing its existential battle because of selling its soul and sacrificing its spiritual direction at the altar of science and pecuniary materialism. The victory of the technological and materialistic society has led to a more selfish and unethical society where individualism, hedonism, individual gratification and aggrandisement are the order of the day. This is why the pro-technology development thesis of this book is not based on the same sentiment as that of atheism. Science and technology can be used for development and national competitiveness without venturing into areas that will redefine what it is to be human in a deleterious, mechanistic manner.

As it should be evident from the foregoing, information technologies have tremendous potential applications and impact on society, including Africa. Consequently, examining the sociocultural implications of IT, particularly exponential technologies in Africa is imperative. Therefore, African countries should explore ways to influence or modify the various technologies emanating from other parts of the world, as much as the technologies developed locally, so that they can fit the different African societies or countries using them.

The Implications of IT on the Sociocultural Realities of Africa

Technology is not just a technical artefact; individuals with a particular worldview create it. Therefore, it carries the culture and values of the people who make it. This means technology is not value-free and has social and cultural components which affect the lives and work processes of individuals, their social relationships, and the religious and cultural views of a people. In addition, science and technology also affect the environment. Consequently, it is imperative to consider the sociocultural aspects of information technology – and more specifically, exponential technologies – when deciding which technology to develop, design, implement, and use.

A society's culture is seen in its citizens' values, customs, and practices. On the one hand, technology dramatically impacts all aspects of culture, including language, arts, religion, education, and so on. On the other hand, a society's culture influences how people perceive, judge and evaluate every technology. For technology to have the desired impact and be entirely accepted by a people, it must be designed to be in tandem with the elements of the culture of the people, such as language, names, music, and others.

In addition, the impact technology has on the environment necessitates that in designing technology systems, the sociocultural elements of the society where the technology will be implemented and used are to be considered. This means that the culture of the environment where the system would be implemented and used should be embedded in systems design, more like the indigenisation of IT. Integrating the culture of an environment in system design offers a high chance of acceptance of innovations and technology within any given society.

Cultural differences between various groups are deeply rooted. As Africans, we are highly invested in our culture because it is the medium through which we live our lives. Moreover, we have languages, values, norms and social practices that are uniquely African. With such uniqueness of norms, values and social behaviour, it is imperative to design technology systems with characteristics adapted to the sociocultural context of the continent.

Imported technology comes with the cultural identities of the countries of origin. IT is laden with the purposes of its designers, but it is also often transformed at the point of use. IT has brought many positive changes, such as improving decision-making, efficiency, effectiveness and competitiveness of organisations, industries and administrations. The exponential technologies are poised to get even more favourable advantages. However, the net effect of these positive impacts could be deleterious if not adequately monitored. The potential for technology to automate the labour process has occasioned an extensive elaboration of the Marxist Labour process theory. Technology plays a vital role in culture, often mediating significant choices and changes, forcing a particular mode of industry, living and perception of reality. Many today have started noticing the importance of IT as part of the contending values of society (Aral, Brynjolfsson, and Alstyne, 2012; Turkle, 2015; Bao, Chang, & Barzilay, 2021).

The clock provided the world with standardised time. Similarly, automation is now leading to the standardisation of information, and artificial intelligence, considered "the grand daddy of all exponential technologies," is striving to standardise knowledge, human intelligence and creativity. Already, some computers tell humans that what they have argued or designed is not valid and why. If AI practitioners can achieve the advances they promise the world, IT will define sound reasoning, inference, or intellectual feat.

How AI Could Threaten the Sociocultural Realities of Africa

As explained in Chapters 4 and 5, artificial intelligence is a technology that seeks to design computer systems that simulate human intelligence or cognitive ability. From a philosophical perspective, two types of AI hypotheses have been described based on the argument of John Searle (Frankish & Ramsey, 2014). These are strong AI and weak AI. On the one hand, the strong AI hypothesis seeks to produce machines with human consciousness and intentionality (emotions, understanding and intelligence). On the other hand, the weak AI hypothesis aims to build machines that only simulate some human intellectual attributes or qualities. In doing so, however, it is to understand those phenomena better and to assist humans in the performance of some task – just as expert systems now aid expert problem-solving and decision-making – without concluding that the AI systems produced to mimic human qualities possess such qualities themselves, like life, intelligence, consciousness, or intentionality. Many want to build systems that can capture human intelligence and eventually take over from human experts in domains of expertise.

Today, there are AI systems that translate languages, communicate with humans, control operations in the industry, and which are obedient servants. In the future, they could surpass humans in tasks of intelligence and creativity. This is a future in which genetic engineering and bio-engineering could create new species of beings that cross between one life form and another and where men could "bear" pregnancy.

In fact, many AI practitioners expect machines to be developed before the end of the 2020s, fully utilising the raw computational power to mimic all that goes on in the human mind. According to Ray Kurzweil, futurist and co-founder (along with Peter Diamandis) of Singularity University, artificial intelligence will achieve human-level intelligence by 2029 (Galeon & Reedy, 2017). Even furthermore, by 2045, according to Kurzweil, all technological developments, especially those related to artificial intelligence, will produce more intelligent robots than humans. He calls this phenomenon the "singularity," describing it as the period when functional human intelligence will combine with computer intelligence to produce a billion times more practical intelligence. Kurzweil says that devices (or chips) will be created in the 2030s to penetrate your brain and enhance memory capacity. He predicts a future of unprecedented human-machine synthesis. In Kurzweil's own words, "Ultimately, it will affect everything. We will be able to meet the physical needs of all humans. We will expand our minds and exemplify these artistic qualities we value" (Galeon & Reedy, 2017).

There could also be reincarnation and rebirth of people in the future. According to Kurzweil (2005), with nanobot mechanisms and scanning technology advancements, humans can send nanobots into people's brains to extract memories of loved ones or deceased people. With enhanced DNA sampling technology, DNA will be collected from people's memory and be able to construct an effective virtual representation of them.

The aspirations of some AI practitioners are overzealous, especially regarding human versus machine consciousness. What is consciousness? As many in the AI world believe, is it merely the physical and chemical phenomena of the human brain? Mbaegbu (2016), Mbiti (1975), and Mbanaya & Cezula (2019) see Africa's perception of consciousness or the mind and body dichotomy to be similar to what obtains in most other parts of the world. They view consciousness as an inexplicable by-product of humanity, as the creation of an intelligent designer, or creator, endowed with spirit, soul, and human intentionality. The primacy of a Creator in creating humans is thus highly revered, as is the spirituality of humans.

Those in the AI world, especially proponents of strong AI hypothesis, who believe machines can be made to be human when appropriately programmed, believe that the world and the nature of humans can be explained through mechanistic reductionism. Thus, they provide a mechanical and reductionist explanation of consciousness as simply a by-product of human brain functions.

However, if this were true, human senses, identity, and personhood would be a delusion. Such reductionism is one of the legacies of 20th-century science.

The strong AI hypothesis sees the human mind as similar to the processes that take place in a computer. To its proponents, a human mind is software run by the human brain (the human "wet-ware"), just as computer software is run by its hardware. Therefore, one can study the brain's software, capture it in a computer programme and then install it in another hardware medium, such as a powerful general-purpose computer, to encode it.

Machines today are much simpler than biological organisms, but they are growing more complex, perhaps a thousand-fold more complex every 20 years. Therefore, within 10 to 20 years, man will have the raw computational power to mimic all that goes on in human minds, perhaps through mighty parallel processing hardware machines; man will be more knowledgeable about capturing human common sense, intentionality and expertise. The robot industry will go out into the solar system and get increasingly sophisticated while humans get increasingly 'useless' here on earth. There will be superhuman machines in a century, and the world will be a backwater wherein biological humans question their essentiality.

To an African mind and sensibility, it sounds quite ridiculous to think of infusing human life into machines, that is, if the agenda of the strong AI hypothesis proponent to create a device with the full complement of human cognitive abilities is achieved. Imagine, as predicted by Kurzweil (2005), computers gain human-level intelligence by 2029; Africans will be living in and sharing a future world with such machines, which will be given citizenship of nations and equal rights with humans. Such Artificial General Intelligence (AGI) systems can also use humans as pets, or subordinate humans, when they become Artificial Super Intelligent (ASI) systems – in other words – as they become superhuman and super-capable. Thus, Africans will live in an order they did not bargain for – a world that goes against some of their critical cherished beliefs and sensibilities. This is why Africa has to participate in the debate concerning the future regarding what types of AI machines should be allowed to be produced.

Despite what some proponents within the AI enterprise think, humans are unique organisms; there is more to a person than mere mechanical processes. One cannot reduce all that constitutes humanity to a machine code. Humans have immaterial values, such as love, joy, peace, patience, commitment, loyalty, etiquette, and morality. These values defy mechanistic, reductionist explanations. They are not tantamount to physical brain functions and, therefore, cannot be captured in another medium by shovelling data and machine code. People are not mere machines. They have free will and unique experiences. Conscious experience is not just a result of the workings of some atoms. People have minds and bodies. The mind cannot be reduced to only the working of the brain. It is absurd to the African sensibility to suggest this.

Although there are particles in the human body, there are also very complex mental phenomena which cannot be reduced to the sum of these particle paths. Indeed, the nature of consciousness is much more than the sum of the particle paths of human bodies. The self and consciousness cannot easily be explained away via reductionism. With the knowledge at our disposal today, it is difficult to use parts of anything, be it atoms or computer codes, to create a fully functioning human being. Many conscious experiences are unique to human beings. One cannot build a human being without understanding the input and output of how neurons affect one another.

It is not likely that people can download human intentionality onto silicon, for even lizards have millions of neurons but are still not considered intelligent in a human sense. The brain creates relations in a way that machines will perhaps never be able to imitate. It is impossible to talk about memory without its subjective aspects. A person's memory cannot replace that of another. Also, *human* perceptions, experiences and memories are embodied; they cannot be devoid of human bodies. People who are partially paralysed and cannot feel some parts of their bodies treat the paralysed legs differently, as if they are not there at all. They often shave one part of their face and leave the other unshaven, feeling that they have shaved all parts of their faces as if the "unfeeling" part does not exist. So, in movements, in explorations of the world of sensation, *human* experiences are embodied. It is through the body that human beings explore and understand the world. Thus, loss of sensation in any part of the body gives an altered image of that part of the body.

When it comes to higher mental processes like consciousness, the brain is unlike a computer. One cannot talk about human knowledge without talking about a "self" integrated into the living body. Every individual explores and categorises experience within a personal centre. Intentionality cannot be mechanised; it requires a living body in a world of affairs and relationships. People do not compute to smell something, dream, or understand a colour. So, at best, computers can only be programmed to simulate different types of human intentionality, but not to be said to possess them in a human sense. This means that I believe in the weak AI hypothesis not the strong AI hypothesis.

The West presents a machine metaphor to the world and humans through computer science, especially AI. Africa needs to participate in the debate about AI (and other fields such as genetic engineering); otherwise, it will find itself willy-nilly living in a new philosophical paradigm that is antithetical to African sensitivities. By participating in this debate, Africa may perhaps sway the world to adopt a more holistic paradigm, assisted by lessons from quantum physics and the theories and speculations of the green movement. For example, quantum physics and the idea of relativity provide a systemic and cybernetic metaphor of the world due to the wavelike, sometimes particle-like, and observer-dependent reality of substances at the subatomic level.

In science fiction (some of which could be called philosophical and futurist literature about AI or machine intelligence), human-made life forms and intelligent machines are recurring themes. While cinematic special effects and CGI have become more sophisticated over the years, the stories remain unhappy. The machines turn around and start behaving in a manner that is unpredictable to humans; they sometimes start persecuting humans, even their creators. The machines or robots claim they are "alive" and, therefore, should have the rights which humans have in society. Then we see humans begin to interact emotionally with machines, empathising, sympathising and befriending them. Some machines cannot understand why some humans treat them as if they are not alive. Many of these science fiction stories are variations of intelligent machines feeling betrayed by humans and vice versa. There are also stories of machines that go mad.

Science fiction is fantastical, but it provides a disturbing revelation of the megalomaniac ambitions of some of the Al intelligentsia, who want to create machines that can surpass humans in feats of intelligence, which have human emotions, communicative patterns and behaviour, and human understanding.

The proponents of strong AI not only want AI to join other machines in the continuing mechanisation of the world but also believe that machines can and should take over from humans in the future, claiming that AI is the following line of evolution. Therefore, they seek machines to surpass human levels of intelligence. AI is a Promethean science, and members of the strong hypothesis school believe it is now the new king of all the sciences, that AI will free the world from anti-scientific dogmas, religion, and supernatural explanations of humanity.

Africa must ask more profound questions, and should we want to create a world where AI does not win the battle for our souls and enslaves us, we must examine and decide the direction in which technology, especially Al, shapes the destiny of the world (including Africa). Many people, even in the West, are offended when confronted with AI. They regard its goal as an affront to their humanity and dignity. Nevertheless, the engine of AI marches on as if it now has a life of its own because the establishment involved in AI does not welcome alternative views or listen to the fears and concerns of the conscientious among them. Many Al proponents cloak the reality of what they are doing from the individuals who will have to live in the society that these technologies will bring about.

The technologies at our disposal today are products of choice. We should therefore ensure that we make the right choices about the technologies we build and how we build them. Unlike the ethics of nuclear technology, the ethics of computer and IT engineering are murky and indistinct. It is possible to make prescriptive generalisations, but they are complex and satisfying. To say, for example, that computer systems may be used to process and provide information without usurping social decision-making is to evade the issue. Technology is the embodiment of specific practices and priorities of culture, ethics and philosophical positions; its dilemmas are resolvable only in particular terms.

AI and computer science create such dilemmas because, despite all the agendas and misunderstandings that suffuse them, they expand the potential scope of automation. The question of what should be automated or who should make the decision becomes crucial. Leaders? The people of the West? The scientists of the West? This is, therefore, not just a question about technology but its social impact, spiritual impact, and political power.

AI and other technologies, such as genetic engineering, will expand the world's automation capability and give us more power to exploit the environment for economic benefit. Africa seeks more power in technology to also be able to exploit its environment for economic benefits. Does this mean that AI enhances this capability, should be applied uncritically or should it be welcomed without circumspection? Societies still have to decide for themselves what to do, for this is not benign technology; it is an empowering, defining and Promethean technology which will bring about varied social, cultural, philosophical and ethical implications and impacts. Its development should, therefore, be proportionally checked with critical and conscious appraisal and a careful decision about its ultimate utility.

The Need for Technology Assessment in Africa

Geneticists, scientific organisations and governments all over the world are forthrightly opposing American organisations attempting to patent fragments of human genes without knowing what they do in the body. For example, in a landmark case in the US in 2013, between the Association for Molecular Pathology and Myriad Genetics Inc., it was ruled that the company should not try to patent any genes, as they are gifts from nature and not what they wholly invented (Justia, 2013). The attempts have also been heavily criticised by Nobel prize-winner James Watson, who resigned as the head of the US government's National Institute of Health (NII-I) human gene research programme. The European Community (EC), Japan, China, the American Society of Human Genetics, the Royal Society and the Human Genome Organisation (HUGO) oppose the patent application because of the ethical and spiritual questions they raise about humanity. They are concerned about reducing human beings to products, tampering with human genetic makeup, and scientists practising "alchemy" on human tissues and genes.

Once again, Africa did not react to this because it did not realise that what is common practice in America today influences the rest of the world tomorrow and can also become standard practice on the continent going forward. Thus, Africa can find itself in a world where certain technologies have engendered new types of reality, cosmology, beliefs and practices they would vehemently have opposed.

This means that Africa and other developing countries should endeavour to understand new technologies' economic, cultural, intellectual, social, philosophical, and workplace implications. Africa must have a centre for monitoring and theorising about the impact of and the influences of new technologies that would define the world's future. People live today in a shared world system in which economic, sociocultural, and political relationships bind together within everyday political, cultural and economic realities. This being the case, Africa should stop ignoring the new scientific developments in the developed countries as something that pertains only to them; the logic and imperatives of everyday existence and relationships between different parts of the world today call for a different stance.

There is a need to analyse IT and the exponential technologies that are being used (or potentially going to be used in the future) in Africa and elsewhere in the world. Technologies used in societies and organisations have many types of impact on, for example, the environment, the nature of work, organisational competition, wealth generation and the level of national development, human lifestyle and the sociocultural life of a people. This is the reason why there is a need for institutions and individuals to be involved in the business of assessing technologies (present and potential). The findings, reports and recommendations from the organisations and individual researchers and analysts should be open for discussion in society. They should inform, for example, the dialogue between management and employee unions concerning what is essential for the welfare of the business, organisation and employees.

It should also influence the leaders, lawmakers, politicians, interest groups, researchers and the general public so that they, too, can be involved in the final decisions that will affect government policy, technology practitioners and research and design decisions.

Technology assessment enables the whole society to assess and determine what the positive opportunities and negative consequences of a given technology or scientific research are or could be. It should allow organisations to have better information and data that their decision-makers can use to make decisions concerning science and technology in their country and determine the stance they should take concerning developments in other parts of the world.

In this way, technologies will be produced that satisfy not only economic and technical needs (by improving productivity and helping to solve those problems that require a technical solution) but also social and cultural needs, and that will lead to the fulfilment of individuals in terms of their innate needs and expectations of what work should do for them. Technology need not make the environmental problems of a country (and of the world) worse, but perhaps, improve them.

Technology assessment would enable a country to make decisions leading to industries, processes and products that cause less pollution. Also, it will allow a country to have the data and information to understand each technology or scientific activity, such as new medical products and treatments and how they affect society, the individuals, the ethics of the community, the religious beliefs of the organisation and individual risks. Therefore, technology assessment will enable Africa to judge the risks a given technology poses against the benefits it would provide and make the best decision for society.

Technology and Choice: How Africa Can Influence the Various Technologies Emanating Across the World

Not all types of technology are inevitable. Technology does not have a life of its own. It results from choices made by people, perhaps made mainly through scientists, technologists, engineers, politicians and business organisations, and sometimes with the involvement of the people at large. It is due to the conscious and unconscious decisions humans make in a society that certain technologies evolve instead of others. It follows that society's technological trajectories and paradigms bear cultural, economic and socio-philosophical imprints and imperatives.

All the technologies now emanating from the West are not automatic artefacts destined to come to the world at certain specified times. They are here because certain people decided that the world should have them. There is also no inevitability concerning the trajectory a particular technology must follow in the decisions concerning how the technology can be designed, which ones should be pursued and what they should look like, or what function they should have. Thus, there is no inevitability to artificial intelligence technologies or a particular growth trajectory. Western scientists are in their laboratories today because of what is funded and not funded and the preferences of those with the power to influence the technologies they are working on towards a particular trajectory and technology paradigm.

Africa should not wait for all kinds of technologies to arrive without partaking in influencing the technologies emanating from other parts of the world. Africa can choose the types of AI systems, genetically-engineered organisms and technologies to be designed and used. The West's economic interests, choices and actions, including the parts that cultural and intellectual imperialism play in underdevelopment, are already affecting Africa.

Meanwhile, Africa does not attempt to study and challenge Western philosophical positions, values, definitions of reality and worldview. Thus, Africa is perceived as having nothing to contribute to world debate, especially in science, technology, genetic engineering, male pregnancy, surrogacy, artificial insemination, artificial intelligence, space exploration and bio-engineering. Africa waits and follows what has been decided by the West: We do not try to evaluate individualism, materialism, genetic engineering, feminism and the classical Western ethos of morality from an African viewpoint.

Africa must attempt to encourage institutions to debate and study the so-called New World paradigms, sciences and technologies and their implications for Africa and the rest of the world. Western science and social science come from a paradigm that is mainly against collective morality, ethics and religion—ingredients that are vital for a stable society, for a good, rich and purposeful existence in this world. These factors make much sense to many non-Westerners; without them, there is emotional, spiritual and personal emptiness and confusion, and society swiftly runs to mindless entropy.

Most Africans still do not realise that Western science and technology are value- and culture-laden, affected by their prejudices and wishes of how things should be. Therefore, the physical and social sciences transported to developing countries are laden with Western culture and values. Western science should not continue to be regarded as faultless, as implicitly good, because it has created a world which is being rigorously criticised (even by Westerners) for its alienation, de-spiritualisation, materialism, mechanism, scientism and emptiness when it comes to looking at such big questions as who we are, where we are going, and what the good life is. Western science and its products and technology have to be questioned by Africa. There should be institutions, agencies and individuals whose business it is to appraise every new scientific enterprise and inform Africa of its effects and concomitants and, therefore, its implications for policies on how society should progress.

Despite the evident importance of IT systems to the economic and industrial development of all nations, African countries should still endeavour to ensure that the systems they use do not destroy the positive aspects of their culture, cosmology, spirituality and social life. After appraising the social and cultural implications of the AI enterprise (as reflected mainly by the views of those belonging to the strong AI hypothesis school), African countries should establish monitoring and assessment centres for information technology systems and exponential technologies. Governments should use the results and reports such centres produce to guide their technology policies and design their systems and technologies.

References

Aral, E. B., & Marshall, V. A. (2012). Information, Technology, and Information Worker Productivity. Information Systems Research 23(3-part-2):849-867.

Bao, Y. , Chang, S. & Barzilay, R. (2021). Learning Stable Classifiers by Transferring Unstable Features. In International Conference on

Machine Learning (ICML), 2021.*1483-1507*

Frankish, K., & Ramsey, W. (Eds.). (2014). *The Cambridge Handbook of Artificial Intelligence*. Cambridge: Cambridge University Press. doi:10.1017/CBO9781139046855

Galeon, D. & Reedy, C. (2017, March 14). *Ray Kurzweil Claims Singularity Will Happen by 2045.* Kurzweilai.Net. https://www.kurzweilai.net/futurism-ray-kurzweil-claims-singularity-will-happen-by-2045

Kurzweil, R. (2005). *The Singularity Is Near.* New York: Viking Books

Mbaegbu, C. C. (2015). A Philosophical Investigation of the Nature of God in Igbo Ontology. Open Journal of Philosophy 5 (2):137–151.

Mbaya, H. H. & Cezula, N. S. 2019. Contribution of John S. Mbiti to the study of African religions and African theology and philosophy. Stellenbosch Theological Journal, 5(3):421-442, doi:10.17570/stj.2019.v5n3.a20.

Mbiti, J. S. (1975). *Introduction to African Religion*. London: Heinemann.

Meyer, S. C. (2021). *Return of the God Hypothesis: Three Scientific Discoveries That Reveal the Mind Behind the Universe* (Kindle Edition, p. 576). HarperOne. https://returnofthegodhypothesis.com/

Justia. (2013, June 13). *Assoc. for Molecular Pathology v. Myriad Genetics, Inc., 569 U.S. 576 (2013). Justia US Supreme Court Center.* Justia Law. https://supreme.justia.com/cases/federal/us/569/576/

Russell, P. (2004). *From science to God: A physicist's journey into the mystery of consciousness*. New World Library. https://books.google.com.ng/books?id=hZJ6AAAAQBAJ

Snyder, Benjamin & Regina Barzilay, "Cross-lingual Propagation for Morphological Analysis," Proceedings of the 23rd Conference of the Association for the Advancement of Artificial Intelligence (AAAI), Chicago, Illinois, pp. 848-854, 2008.

Turkle, S. (2015). *Reclaiming Conversation: The Power of Talk in a Digital Age.* New York, Penguin Press.

Chapter 11

Making Africa an IT Power House

As I have previously pointed out, the experience of the developed world has proven that technological transformation leads to faster economic growth and job creation. In fact, economists widely agree that technology is the driving factor behind economic growth in countries, regions, and cities. Advances in technology usually lead to more efficient production of goods and services of higher quality, which is the bedrock of the prosperity of any society.

In Africa, there has been a severe lag in technology development, leading to a lack of economic growth across the continent. While the continent is being closely watched as the next development frontier, given that it is home to a vibrant, young population (it is a promising primary consumer market over the next three decades, and it is experiencing an increasing mobile phone penetration), economic growth across African countries has stalled.

Economic development in Sub-Saharan Africa (SSA) is estimated to have decelerated from 2.5% in 2017 to 2.3% in 2018. In the same year, 2018, the real economic growth across Africa was 2.3% (Zeufack, 2019). The IMF and the World Bank revised their economic growth forecasts for SSA downward in their 2019 reports to 3.5% and 2.8%, respectively. According to Chakravorti & Chaturvedi (2019), as of 2019, 437 million of the world's extremely poor resided in SSA, while 10 of the 19 most unequal countries are found within the region.

It is not farfetched to see why Africa lags behind the developed world in technological advancement; it is mainly due to lack of relevant technologies, as well as technological know-how and know-why. For example, compared to the rest of the world, the quality of

Internet access in Africa is inadequate. In 2019, Africa had 39.6% Internet penetration, compared to 62.7% globally (Ghanem, 2020). How widespread Internet connectivity is in a country can also indicate how high the GDP of that country would be. In other words, as an example of technology, the Internet diffusion level in a country also shows how well that country is faring developmentally. In other words, developing countries in Africa spend more to subscribe to the Internet than the rest of the world.

Africa consumed only 1% of the total international Internet bandwidth available in 2017, while the monthly fixed broadband connection cost was 36.6% of gross national income, compared with 14.5% globally in 2019 (Ghanem, 2020).

With such sobering statistics, African countries must give attention to the development of technology to leapfrog ahead in the continent's economic growth. Digital technology, spearheaded by exponential technologies, offers African countries a rare opportunity to overhaul various sectors of their economies.

It is encouraging to note that many African countries are taking steps toward complete digitisation, which is a significant step toward turning the continent into an IT powerhouse (Ahmed, 2020; Bayuo & Bamford, 2021). Over the last two decades, Africa has had the world's highest Internet usage and connection rate. During the same period, Africa's Internet user base grew from 4.5 million to 523 million (Ghanem, 2020), while in the rest of the world, the increase was hardly consequential. So, even though Africa is growing in Internet use from a low base, it is rapidly increasing access to the Internet. This is encouraging for Africa and should be further increased or sustained for equality with the rest of the world.

Across the continent, from Nigeria to Kenya and South Africa to Egypt, young Africans are capitalising on the new technologies penetrating the continent to launch successful technology startups and innovations that have already started to add to the continent's economy significantly.

In Kenya, Chura, a web and SMS-based service, allows mobile users to access airtime and mobile money services across different mobile networks in the east African country. In Nigeria, two leading fintech companies, Flutterwave and Paystack, have revolutionised online payment. Flutterwave, a fintech company which helps customers send, receive money and pay bills across Africa, has processed 100 million transactions. Paystack's online payment gateway enables small businesses and individual entrepreneurs to create pages and receive payments from customers using Visa cards, Verve cards or Master Cards from any part of the world.

In Rwanda, Zipline delivers medical supplies using automated drones, while Nilebot, an Egyptian agritech startup, is an IoT-based aquaculture water quality monitoring system. The system includes integrated sensors that measure water characteristics, including temperature, oxygen saturation, dissolved oxygen, pH, salinity, and total dissolved solids.

With the efforts of young Africans demonstrating the potential of digital technology to help Africa leapfrog its development, African countries must pursue an initiative for digital transformation to turn Africa into an information technology powerhouse. Such an initiative should include elements such as:

1. Ensure Top-Notch IT Education in Primary and Secondary Schools

A "bottom-up" investment in IT education is one of the best ways to make an African IT powerhouse. By this, I mean that the continent should invest in IT education from primary to secondary school. The goal is to identify potential as they are still learning to use computers. However, to achieve this aim, some initiatives and actions need to be taken as follows:

- **Bottom-up STEM education**
- **STEM/IT-savvy teachers**
- **Providing African primary/secondary schools with IT equipment**
- **Continental/Regional IT education alliance**
- **Government/Non-government partnership**

Firstly, African countries should overhaul their educational curricula to reflect the IT needs of the continent. Unfortunately, due to colonial factors, the education systems in African countries lack the requisite tools to make IT excel. African countries must develop and apply science, technology, engineering, and mathematics (STEM) education to correct the setback.

Can STEM education in primary and secondary schools be possible without STEM-savvy teachers? No. In that case, it is not enough for the governments of African countries to initiate and implement STEM education right from primary school to secondary. It is critical to ensure that teachers at these levels of education are capable of teaching STEM. Unfortunately, we find scenarios where many teachers in primary and secondary schools across Africa are not adequately learned in STEM education, so they cannot give what they do not have. Getting STEM teachers ready for the IT needs of African children falls squarely on the shoulders of African governments and other relevant stakeholders (STEMpedia, 2019).

To get IT/STEM-savvy teachers for African schools, African governments can re-train their available primary and secondary teachers to become IT-equipped. However, achieving this feat will not be easy because most available teachers might not easily catch up due to age and other factors. So, the next option could be for African governments to recruit young and vibrant teachers and establish special STEM/IT programmes for them that can last three years. With this IT training, it will be easier for such teachers to impart IT education to African children.

However, the government's initiative and implementation of STEM education and re-training and employment of teachers is not enough; it is like sending farmers to farms without the necessary implements. Therefore, African governments should equip primary and secondary schools with STEM/IT-enabling tools and equipment, such as modern computers, strong Internet connections, software, IT assistants and engineers, etc. Providing IT tools for pupils and students alike will help them become IT-compliant because one of the best ways to learn is to participate. With IT equipment available in their schools, it will be easier for them to acquire basic IT knowledge at a young age.

Apart from providing primary and secondary schools with IT equipment, Africa can achieve ground-breaking IT education for African children through continental and regional IT education alliances and cooperation. This would entail the efforts of different federal ministries of education in various African countries, the AU, and other African regional bodies, such as the East African

Community (EAC), Southern African Development Community (SADC), Economic Community of Central African States (ECCAS), and Economic Community of West African States (ECOWAS). These stakeholders can formulate and implement IT education programmes for African children. Training, re-training, and teacher-exchange programmes could be part of the continental and regional IT education programmes. These bodies could come together to set up trust funds for grants so that Africans who desire to become researchers or pursue PhDs in areas with high-end research capabilities can be supported to the maximum to increase the likelihood of some of them becoming outstanding researchers in IT education for primary and secondary schools.

African governments cannot achieve top-notch IT education for children on the continent alone. This is where the efforts and collaboration with regional, continental and international partnerships become essential. Therefore, Africa should partner with UNESCO, AfDB, regional and continental organisations, companies, and others. The aim of the association should be to equip African primary and secondary schools with human and non-human IT resources.

In the last analysis, following the steps above will help Africa solve many IT challenges. This belief is simple: innovation requires many clever and highly skilled tinkerers. The larger the number of children exposed to high-quality IT education at the primary and secondary levels, the higher the chances of some of them becoming the next generation of IT experts who will change the course of events through creative innovation.

2. Invest in the Development of a Digital Savvy Workforce in Africa

From **Figure 27**, it can be established that digital skills in Africa need urgent attention. The vast digital skills gap is one of the factors diluting economic opportunities and development on the continent (World Bank, 2021). So we are now in the digital era, where digital skills are some of the most dependable ways to get work everywhere in the world and where 230 million jobs in Africa will require some level of digital proficiency by 2030 (according to the International Finance Corporation, IFC, 2019). Relevant stakeholders in Africa should aggressively invest in the digital skills of Africa's workforce.

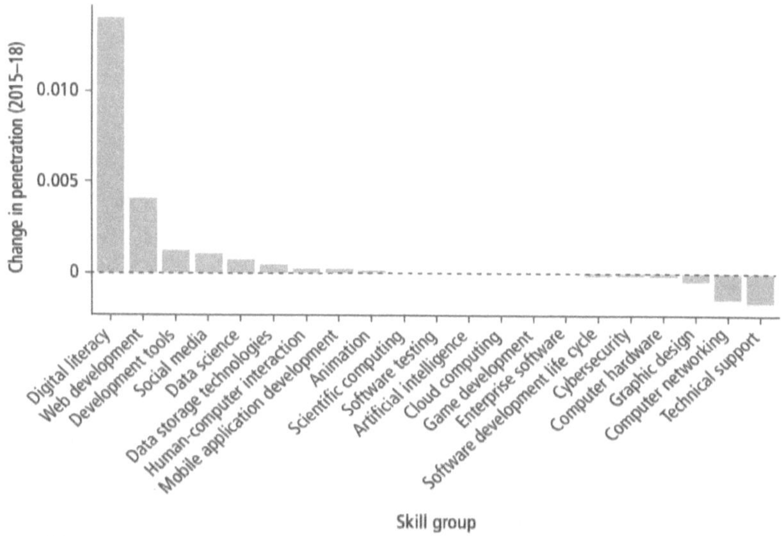

Figure 27: Growth of Digital Skills in Sub-Saharan Africa
Source: World Economic Forum

These are the steps to boost the digital skills of Africa's workforce:

Invest in IT education in higher education: Most African workforce are graduates from different institutions: colleges of education, polytechnics, and universities. Unfortunately, many take jobs every year without the necessary digital skills. Why? Because they are not digitally trained in their schools, and when they join the workforce, their lack of digital skills upsets the economies of African countries; their input falls below the standards of the digital economy. So, tertiary institutions are crucial to investing in the digital skills of Africa's workforce (Fredua-Kwarteng, 2021; Caballero & Bashir, 2020).

To keep up with the rapid technological advancements, tertiary education institutions on the continent must modernise their ICT and engineering curricula. Also, introducing quick skilling programmes is another approach to accommodate the surge in demand for specialised digital needs and skills.

Providing access to digital infrastructures: The application of digital skills can only complement the requisite digital infrastructures. The digital infrastructures include affordable computers, helpful software, electricity, and wide broadband Internet connection. Unfortunately, there is a significant gap in the provision of such IT enablers on the continent on a large scale, but with only considerable infrastructure in centralised settlements. Only 35% of Nigeria's schools are connected to a power source; less than half of the populations of Cote d'Ivoire and Rwanda have access to broadband, which is a problem; less than 25% of people in Mozambique have access to broadband connectivity (Caballero & Bashir, 2020). Also, basic digital skills in Africa are limited by high prices for data, lack of basic literacy, and limited availability of content in local languages. With these digital infrastructural deficits, it is hard for Africa's workforce to be digitally relevant.

Making IT skills everybody's business: We have seen that IT is a tool that drives economic development in all sectors of the economy. Therefore, making the African workforce digitally savvy is providing the tools, environment, and training to help them apply IT in their respective areas. According to Caballero & Bashir (2020), as more organisations use digital technology in various industries, the majority of demand for digital skills will come from generic vocations rather than precisely defined ICT professions. Basic digital skills will account for about 70% of orders, with non-ICT transitional digital skills accounting for the remaining 23% (International Finance Corporation, 2019). So, irrespective of one's profession, the implication is that one needs IT skills. For example, managers in the agricultural industry might need training in using financial software to track income and spending. Likewise, office workers must be familiar with cutting-edge software for creating online presentations. Finally, tourism companies may want to investigate digital marketing tactics that might help them develop their company, such as which social media platforms to use and how to create aesthetically attractive websites. The main implication is that millions of African workforce will require digital skills training or re-training.

The overall benefits of digital investment in Africa's workforce

With the relevant digital skills, Africa's workforce will be helpful in the present and the future. Many jobs today require digital skills. Similarly, as the COVID-19 epidemic forces more organisations to become digital to survive, the need for digital expertise will continue to rise. This forecast is consistent with preliminary findings from the

International Finance Corporation (2019) and the World Bank (via the Digital Development Program Trust Fund): by 2030, significant levels of digital skills will be required for 50-55% of jobs in Kenya, 35-45% in Côte d'Ivoire, Nigeria, and Rwanda, and 20-25% in Mozambique. Overall, having a technologically literate workforce is one of the surest strategies for African countries to improve many areas of their economies significantly.

IT tools and equipment do not operate in a vacuum; Africa needs a digital-savvy workforce to realise the dream of making the continent an IT powerhouse.

African governments must continue to address the skills gap that covers all demographics as Sub-Saharan Africa attempts to enhance innovation, adopt cutting-edge technology, and disrupt business-as-usual traditions (World Bank, 2021). In achieving robust economic growth, it is necessary to invest in core skills for the whole population, not just future generations. Unfortunately, far too many youths in Sub-Saharan Africa leave school without the fundamental skills needed to advance.

3. Promoting IT Research and Development (IT R&D) in African Universities

The university is the apex of research and development (R&D). In universities, many scholars, teachers, researchers, fellows, etc., engage in research, dissemination, advancement, and application of knowledge in different areas of human endeavour, all for the service of society at the local, regional and international levels. In the first world economies, R&D is usually based on the perceived needs of their economies. Therefore, based on the relevance of IT in the 4IR, governments and other relevant bodies need to promote IT R&D in Africa's universities. Promoting IT R&D in Africa's universities can be done by:

Strengthening the education system first: While IT is one of the powerful tools for socioeconomic development, it should be noted that IT alone does not solve all development challenges. In the same vein, the university is a system comprising different but complementary parts. So, it is wrong to emphasise IT R&D alone without addressing first things first.

The first thing? The African university system is fraught with numerous challenges, such as outdated and colonial-structured curricula, poor budgetary allocation for university education, lack of a conducive environment for research and learning, inadequate remuneration for lecturers and other workers, industrial actions, and many more. For instance, how can you preach IT R&D to inadequately paid lecturers whose meagre salaries have not been paid? These issues must be addressed to enable IT R&D in Africa's universities (Midega et al., 2021). So, it is a waste of resources to funnel large sums of investment into IT initiatives without complementary investments in revamping Africa's university education.

Creating national, regional, and continental IT R&D across African universities: It is widely agreed that all African countries should have national R&D. However, consequent upon the relevance of IT in the continental development of Africa's economies, it becomes necessary to establish special IT R&Ds on the continent (Fredua-Kwarteng, 2021). The establishment of special IT R&D will be more operational at the local, national, and international levels. At the national level, IT R&D in universities will identify the areas of need in the IT sector. Then at local and international levels, different IT researchers and developers can come together to formulate an IT blueprint for the socioeconomic development of Africa.

Providing IT R&D infrastructures: One thing is to have researchers, but another is to provide them with the necessary tools and equipment. University researchers are not magicians; until African governments provide them with IT infrastructures, they cannot carry out IT R&D. So, the government should give those infrastructures that can enable university researchers in the IT sector to do their research. Those IT infrastructures include stable electricity, Internet connectivity, vast broadband, adequate technical support, etc.

Sponsoring IT R&D in African universities: IT R&D is capital intensive; therefore, independent research by university lecturers across the continent cannot accomplish much with a budget shortage. This is why the governments of African countries need to sponsor IT R&D, right from the national, regional, to continental levels. Supporting IT R&D in Africa's universities has to start with relevant national, regional, and continental IT policies. African leaders ought to come together to formulate IT policies unique to the continent's socioeconomic challenges. Then, the governments can task university scholars and researchers across Africa to carry out IT researches that are tailor-made to Africa's digital needs. The governments can sponsor IT R&D in Africa's universities through special IT trust funds, grants, etc, for the researchers.

Implementing the results of IT R&D from African universities: Africa is known for abandoned projects. So, spending fortunes on IT R&D when the results end in the library and on the Internet is an aberration. I am not saying that all results from IT R&D are implementable; I am saying that results proven to help solve Africa's socioeconomic challenges should not be abandoned by political or ethnic biases.

4. Setting up IT Bricolage Centres & IT Startup Ecosystems

Bricolage centres and startup ecosystems are essential if Africa becomes an IT powerhouse. Therefore, this section examines the concepts of bricolage and ecosystems to see how they can help the continent in IT.

Setting up IT Bricolage Centres

The French anthropologist, Levi-Strauss (1968), introduced the concept of bricolage in his book, *The Savage Mind*. The underlying idea is do-it-yourself entrepreneurship, in which societies build unique solutions by utilising resources already available in their collective social awareness. The Bricolage theory explains how entrepreneurship emerges in economically depressed areas with little to no resources. The philosophy is based on the concept of making something from nothing. Due to the continent's economic and digital backwardness, establishing bricolage centres can be helpful. Let us look at the "whys and hows" for the bricolage centres on the continent.

(i) The Why of Bricolage Centres in Africa

Many works have stressed that Africa lacks the infrastructure and equipment needed for IT growth. In that case, the bricolage system will supplement the IT efforts of the continent. IT bricolage will enable Africans to utilise the few available IT resources to imagine, tinker, and improvise unexpected specific solutions to some of the global IT challenges.

Besides, since the IT sector is more practical than theoretical, embracing the bricolage system will be more effective for the IT needs of the continent. For example, on the Internet, bricolage, rather than rational and organic planning and design, development of a substantial amount of software, some aspects of biotechnologies, and other major technological sectors, would benefit from the availability of bricolage centres. The development of software is an intriguing example of bricolage. Unlike proprietary software, "open source" software is rectified, updated, revised, debugged, and somewhat rewritten by users to match their specific requirements. Sometimes, the most effective software is the result of such modifications.

Another reason for advocating IT bricolage for Africa is its young population strength. Africa, as of today, has the most significant number of youths in the world. There is an added advantage for the continent in that direction. Since Africa has the most prominent young population, the continent will benefit more when the youth are given the required environment for IT bricolage, as they are more adaptive and have shown to understand new technologies better. However, that is only feasible if plans for bricolage centres in Africa are in place.

(ii) The How of Bricolage Centres in Africa

The truth is that the measurement of the success of IT in a country is not based on research but on what tech-savvy and often-overlooked technicians do. In that case, Africa should encourage IT bricolage. To begin with, we must make policies that are IT bricolage-friendly.

However, making policies is not enough. There should be vigorous implementation. One of the ways to implement IT bricolage is by establishing bricolage centres in Africa, starting from national, regional, to continental levels.

In these centres, African governments can provide other incentives for IT bricoleurs to function. The "expert bricoleurs" will teach the young African population IT skills and assist them in experimenting and tinkering, thereby facilitating bricolage inventions and innovations.

Setting Up Tech Startup Ecosystems

By startup tech ecosystem, I mean the interconnected factors and tools that encourage Africans to set up new technologies. Examples of tech ecosystems include universities, funding organisations, support organisations (like incubators and accelerators.), research organisations, service provider organisations, large corporations, and others.

These systems cannot function without the support of African governments. Therefore, for Africa to become an IT powerhouse, African states and governments need to create favourable environments for IT startups on the continent. Building a suitable environment for the IT ecosystem on the continent starts with formulating and implementing good IT policies to provide IT infrastructures, like stable power supply, wide broadband Internet connection, affordable data, tech hubs, etc (AF Bureau, 2021;Bayuo& Bamford, 2021). In terms of tech hubs, there are a number of them in cities across countries such as Kenya, South Africa, Nigeria, and Egypt. However, Africa can strengthen its IT ecosystems by building and equipping regional and continental IT hubs across the continent.

There should be IT startup funding besides building continental IT hubs and centres. Many young Africans have many creative and innovative IT ideas that need funding. Unfortunately, African leaders do not usually give them the necessary support in terms of funding. Instead, multinational companies and countries from First World economies take up such financing and enjoy the benefits. However, if African governments and tech stakeholders can develop and maintain a robust IT ecosystem, in that case, it will lead to African countries climbing the digital maturity scale, greater digital literacy, innovations, and massive GDP increases.

5. Promoting IT R&D in the Private and Public Sectors

In developed economies, it is easier for the government to drive R&D with dynamic and practical policies and programmes and commensurate implementation strategies that birth innovations. However, it is hard for governments to spearhead such R&D breakthroughs in emerging economies like African countries. Africa faces a labour shortage, geographically dispersed national markets, a growing middle class with more discerning spending habits, and a wealth of underutilised or poorly utilised natural resources. These are some of the qualities that create opportunities for private firms of all sorts to promote technology and drive prosperity. In this situation, private firms do not have an option but to become part of the drivers of technological innovations. Private sector participation in tech innovations in Africa will bring many benefits, but African governments must promote it and create the enabling environment for it to thrive.

• **Promoting private sector participation in the tech sector**

It is agreed that private sector participation in IT R&D is necessary for Africa, but there are facilitations that African states and governments can bring to the table. Africa can promote private sector IT R&D through:

(i) **Friendly IT policies**

Firstly, African governments should formulate and implement private IT R&D-friendly policies that can promote private sector innovations in the IT sector. For instance, such procedures should develop and enforce rule-based institutional frameworks that promote property rights, standards, and incentives for risk-taking.

(ii) **Creating national, regional, and continental economies of scales**

By national, regional, and continental economies of scale, I mean making provisions for Africa's cost advantages to emerge at national, regional, and continental levels. In Africa, there are numerous small, as well as contiguous, markets. African governments can integrate neighbouring markets by negotiating trade agreements to create national/regional/continental economies of scale in private sector IT R&D.

(iii) **Creating incentives for private sector participation in IT R&D**

Some incentives can prosper private sector IT R&D in Africa. Such incentives include sponsorships, grants, and loans. The truth is that many private organisations and businesses on the continent have IT-savvy individuals with innovative ideas. However, they cannot bring their tech ideas to life because they lack funds or support. That is why African countries should pool resources to enable the private sector with innovative tech ideas to develop them through their relevant regional and continental bodies like

ECOWAS, SADC, AU, and AfDB.

(iv) **Providing IT R&D infrastructures**

Private sector participation in IT R&D can only be possible if the necessary infrastructures are provided. Unfortunately, private sector IT R&D participants cannot give up or allow general use of those infrastructures independently because they are capital intensive. Therefore, the burden lies on African states and governments to provide these IT R&D-enabled infrastructures, such as IT R&D centres across the continent, affordable computers, helpful software, electricity, and wide broadband Internet connection.

The benefit of private sector participation in Africa's tech sector: The tech innovations created by private sector participation in IT R&D will lead to growth and economic development across the continent. It will also enhance labour productivity and reduce poverty.

- **Promoting public sector participation in IT R&D**

There are measures that African governments can adopt to encourage public sector participation in IT R&D. The steps are as follows:

(i) **IT R&D funding**

Countries in Africa have neglected IT R&D funding. For instance, R&D received only 0.42% of global GDP in R&D funding as of 2019. The overall average stood at 1.7% (Kyobutungi, 2021). So, the continent has to provide IT R&D funding to become an IT powerhouse. The AU and its development agency, NEPAD, may do this by prioritising collaboration with countries to deliver their pledges on IT R&D spending.

(ii) **Partnering with African IT researchers and experts**

Africa has peculiar IT needs and issues as a continent on the development periphery. Therefore, African countries need to work with people who understand the IT terrains of the continent, and the best group are local IT experts and researchers. The local IT researchers and experts can be recruited from universities, research centres, and innovation and technical centres across Africa. Furthermore, African nations can seek guidance and collaborate with organisations such as the Consortium for Advanced Research Training in Africa, the Pan African University, the Alliance for

Accelerating Excellence in Science in Africa, the African Research Universities Alliance, the Coalition for African Research and Innovation, the Africa Centres of Excellence project, and others. Partnering with them in IT R&D helps solve the tractable IT challenges bedevilling the continent but encourages public sector participation.

Our primary focus must be on a new future, and we must urge for increased South-South cooperation and initiatives that benefit the African bloc. African researchers can no longer rely on financing channels that can be cut off quickly, leaving them helpless.

(iii) **Diffusion of IT R&D**

From all indications, IT is a sector that permeates other sectors. Therefore, IT R&D should be diffused among all the sectors of Africa's economies. Usually, IT R&D is fused with national R&D or other relevant ministries. The implication is that public sector participation will be limited. However, if African countries encourage IT R&D in different sectors of their economies, there will be an increase in public sector participation.

The benefits of public sector participation in IT R&D: It will help the continent leapfrog economic development. By leapfrogging Africa's economies, IT R&D will help the continent to become relevant in the comity of continents.

Vibrant public sector participation in IT R&D in Africa will relieve the continent of dependence, making Africa a key player and contributor to world affairs. The COVID-19 pandemic has shown us the terrible dimension of Africa's dependence on First World economies. African nations were among the last to receive test kits, personal protective equipment, and immunisations. Due to this reliance, inventories are restricted, and there are various bottlenecks.

6. Invest in Digital Infrastructures for Africa

Digital infrastructure refers to the physical resources that enable the use of data, computerised devices, procedures, systems, and processes. To empower African businesses to compete in the Fourth Industrial Revolution and keep up with cutting-edge technological innovation, African countries must identify the digital infrastructural deficit they suffer and make concerted efforts to bridge the gaps. After placing them, the next thing is investing in digital infrastructures to improve access and quality. Specifically, the governments of African countries should look to develop the following digital infrastructures.

i. Internet connectivity

Africa needs stable and affordable Internet. Internet connectivity will facilitate the successful operation of the Internet of Things (such as interconnected items, gadgets, computers and mobile devices) and data repositories (such as data centres and clouds) on the continent.

ii. Wide Broadband

One of the things that put the first world economies at the centre is wide broadband availability. To make meaningful progress in this Internet age, Africa should provide vast broadband on the continent. High-quality fibre optics are needed for data centres, fintech industries, digital healthcare, e-commerce, etc. These industries need vast broadband to capture, utilise, and transmit massive volumes of data. With vast broadband on the continent, it will be easier to connect cities, regions, and the entire continent.

The lack of fixed broadband is a central issue in many African countries. The countries operate below the 20% critical mass necessary to achieve the efficient flow of information to boost innovation and economic development (Abdella, 2020).

iii. Communication Satellite

These are satellites that provide network or information services. They use a wide range of radio and microwave frequencies to relay signals, providing connectivity between geographically distant points.

iv. Network Infrastructure

These comprise the hardware, software, network management, and connection tools that support business operations, communication, and network or Internet connectivity. They include servers, wireless routers, Wi-Fi, and others.

v. Data Centre

A data centre consolidates an organisation's central IT operations and equipment to ease the storage, processing, and distribution of data and applications.

vi. Cloud Computing

Cloud services are provided and paid for to assist organisations in lowering operational expenses and in running their infrastructure more productively. Cloud computing services contain servers, software, networking, analytics, intelligence, storage, and databases.

vii. Platform Category

Digital platforms have a wide-ranging application and enable much of the digital economy. They are based in the cloud and usually offer such benefits as reducing barriers to entry, reducing fixed costs, reducing prices, and making it easier for digital businesses to gain customers and reduce transaction costs.

viii. System Software

System software provides a platform on which application software runs. It is used for running computers. It operates in the background to maintain the computer's essential functions, allowing users to run higher-level software applications for specific tasks.

ix. Application Software

Application software is individually designed as tools to enable the end users to perform particular processes that may enhance productivity or communication.

x. API and Integration

An API is a programming interface that allows various platforms, systems, and applications to collaborate and share information. It will enable the platform to process requests seamlessly and delivers data to facilitate connectivity between devices and programmes.

xi. User Equipment

These devices are designed for end users for communication, such as mobile phones and laptops equipped with a mobile broadband adapter.

7. Invest in the Development of Digital Financial Services

Digital financial services (DFS) comprise payments, remittances, and credits accessed and delivered through digital channels, such as mobile devices, the Internet, etc. The enormous increase in registered mobile money accounts shows that DFS is becoming the primary driver of financial inclusion and wealth creation. Over 1 billion reported mobile money accounts were recorded globally in 2020, with more than half in Africa (Awanis, Lowe, Andersson-Manjang &

Lindsey, 2021). DFS lowers costs by taking advantage of economies of scale while maintaining transaction speed, confidentiality, and accountability. As a result of the impact of DFS on Africa's economies, African countries need to invest in its development. Africa can invest in the development of DFS in the following ways:

i. Formulate conducive legal and regulatory policies and frameworks

To allow mobile phone operators to provide financial services, modifications to monetary and economic rules, alongside control mechanisms, are required to develop financial systems. A digital financial services policy that permits non-bank e-money insurance, adopts simplified Customer Due Diligence CCD, and establishes a broader agent network will all assist in broadening access. Furthermore, African countries should follow the World Bank's recommendations: adopt digital payments legislation; allow non-banks access to payment systems; establish a robust consumer protection framework; create and enforce a competition policy; establish a comprehensive regulatory scheme for DFS providers; adopt a broad legal indicator for privacy and data protection (Pazarbasiogluet al., 2020).

ii. National, regional, and continental support systems & incentives

For DFS to thrive on the continent, Africa should create and provide a financial management system and incentives to support the intensive shift of government-to-person (G2P) payments to digital at national, regional, and continental levels. Africa should also establish and expand coverage of digital ID to achieve more inclusive digital financial services. African governments can encourage DFS by funding some programmes to facilitate its growth on the continent. One typical example is the Africa Digital Financial Inclusion Facility (ADFI), a special fund with a pan-African mandate from the AfDB. Through strategic and catalytic investments across Africa's DFS ecosystem, it addresses the structural barriers preventing the growth and adoption of digital financial services.

iii. Providing DFS infrastructures

DFS cannot operate effectively without the appropriate

infrastructure on the continent. In order to achieve high smartphone penetration, African governments must support widespread broadband connectivity, well-functioning payment systems, and interoperability. They must also build credit infrastructures and expand the coverage of credit-relevant data.

8. Promoting the Mastery of Exponential Technologies

Exponential technologies are those technologies that bring change and economic development to economies at an accelerated speed. The exponential technologies of the 4IR include Artificial Intelligence, Game Design, Robotics, 3D Printing, Augmented & Virtual Reality, Machine Learning, and others. To be part of the digital world and benefit immensely from the 4IR technologies, Africans need to have mastery of the exponential technologies. Therefore, African governments and other relevant stakeholders on the continent need to do the following:

i. **Overhauling Africa's educational curricula to accommodate digital education**

   The exponential technologies of the 4IR have brought dramatic changes to all fields of human endeavour, including education. Therefore, there are emerging skills that define the current global job market. However, as a continent of developing economies, Africa has to overhaul its education system to accommodate the emerging abilities of the 4IR.

ii. **Providing infrastructures that aid the acquisition of exponential technologies**

   The African continent is home to an infrastructural deficit in exponential technologies. The lack of these infrastructures hinders the mastery of the exponential technologies of the 4IR. Therefore, for Africans to master the exponential technologies, the African governments – from national, regional, to continental levels – should provide those infrastructures, like stable and constant power, affordable data, affordable computers and smartphones, Internet connectivity, and vast broadband.

iii. **Recognising certifications acquired from online learning platforms**

   Numerous online platforms offer tech education in the different areas of the 4IR. Some notable ones include Coursera, Udemy, FutureLearn, and BeSingular. These online education platforms cover programming, iOS app development, 3D printing, Internet of Things (IoT), and Augmented & Virtual Reality (AR/VR).

However, traditional and civil service job givers are still cynical about certifications from those platforms. For many, the classes are not as rigorous and complex as those obtained through traditional education curricula in Africa. This attitude douses the enthusiasm of African youths for learning on such platforms. Therefore, African employers of labour who frown at digital certifications should understand that in the 4IR, learning does not necessarily need to be challenging and frustrating as obtainable in traditional classrooms; what matters is that a learner can acquire and utilise the requisite skills.

iv. African youths embrace digital learning themselves

The truth is that the government cannot do everything for everybody simultaneously. There are things Africans can do by themselves. So, African youths should take it upon themselves to learn the emerging digital skills of the 4IR. The willingness and zeal should be high among African youths because acquiring digital skills is one of the surest ways to remain relevant in the 4IR. With these skills, African youths can stay in their remote locations and get international jobs online, as far as they are connected to the Internet.

9. Formulate Continental Policies for the Technological Unity of Africa

The development of the first world economies did not happen by accident. Their economies follow a unified and interconnected developmental blueprint. Therefore, for Africa to leapfrog the development of its economies with the help of the 4IR, there ought to be technological unity among African countries. The following steps would be helpful:

i. Identifying the technological barriers bedevilling the continent

It is easier to solve challenges when they are identified. In Africa, numerous challenges hinder the technological unity of the continent and prevent the continent from developing economically. That is where Africa needs the services and opinions of IT R&D experts and scholars. The continent should task IT and R&D experts with identifying Africa's peculiar technological barriers at national, regional, and continental levels. Once that is done, the continent may unite to address these issues.

ii. Continental cooperation

Formulating continental policies for the technological unity of Africa starts with continental cooperation. African policymakers must collaborate at the regional and national levels. Strategies for national digitalisation cannot be executed in isolation. The African Continental Free Trade Area demands international collaboration in digital taxes, data security, privacy standards, cross-border data exchange, and interoperability to link the continent's digitalised economy.

iii. Implementation is paramount

Today, there are worthy technological policies in Africa. Nevertheless, as a continent, executing well-deserved policies is always a significant problem. So, it is not enough to make continental policies on the technological unity of Africa; African governments need the will and practical demonstration of their implementation.

10. Ensure Large African Content in Digital Technologies

One of the best things the 4IR offers to Third World economies like Africa is exponential technologies. With such technologies and the necessary efforts, it is possible, though tasking, for developing continents to leapfrog and become global players in the digital technology arena. So, in this section, I argue that Africa has the prospects and potential of having significant African content in global digital technologies if they utilise the following indices:

i. Large global youngest population

The young population of the first world economies is decreasing. Fortunately, the African youth population is increasing. More than 60% of Africa's population is under 25, with a third between 15 and 35. So, with a median age of 35, Africa will still boast the world's youngest population by 2100 (Rocca, 2019). Africa's most youthful population implies that digital technologies are more familiar to the younger demographic. Invariably, with the majority of Africa's youth population, the continent can become part of the global tech players in the future. Despite these statistics, African markets remain one of the most significant sources of untapped young potential all over the world, so the continent can only become a global tech player if it harnesses the potential of its young population.

ii. Large consumer market

Africa's consumer market is among the biggest and fastest-growing in the world. The continent's consumer spending has increased at an annual compounded growth rate of 3.9% since 2010, realising $1.4 trillion in 2015. This amount is predicted to rise to $2.1 trillion by 2025 and $2.5 trillion by 2030 (Signé, 2018). Furthermore, if the Continental Free Trade Agreement (CFTA) is effectively employed, a unified continental market for products and services will be effective by 2030, giving firms various access points to the region and a prospective market of 1.7 billion people (Signé, 2018).

These reports have implications for the continent's position in tech globally. The studies give an overview of Africa's most significant potential in the digital consumer market sector to tech investors, startups, and business leaders, among others. The metrics also provide African tech policymakers with information and perspectives on the options most likely to attract private investors, accelerate consumer market expansion, and contribute to economic growth and poverty reduction, thereby positioning Africa as a critical player in the world's tech industries.

iii. Address the current digital hitches confronting Africa

Having significant African content in the comity of global digital technologies is impossible without addressing some of the harsh realities confronting the continent. Firstly, the continent needs to address poverty. In Sub-Saharan Africa, the number of people living in severe poverty is increasing, representing more than half of the extreme poor in 2015. Studies show that Sub-Saharan Africa will be home to nearly 9 out of every 10 impoverished people by 2030 (Wadhwa, 2018).

Likewise, Africa is still catching up to global standards in terms of digitalisation. Internet prevalence and speed are relatively low, and the cost is high compared to the rest of the world. In 2019, Internet penetration in Africa averaged 39.6%, compared to 62.7% globally, with substantial variances between nations, ranging from 89.8% in Kenya to 5.3% in Burundi. Furthermore, Africa achieved 21% growth in mobile broadband network (4G) rollout in 2020, while growth in all other regions was negligible (International Telecommunication Union, 2020); and the median mobile broadband download speed is 2.7 Mbits/s, nearly half the global average of 5.2 Mbits/s. The monthly expenditure of a fixed broadband connection is projected to be 36.6% of the total national income, compared to 14.5% internationally (Ghanem, 2020). Digital investment accounts for roughly 1.1% of GDP in African countries, compared to 3.2% in advanced economies. As a result, continuing business as usual is no longer viable, as it will exacerbate the digital divide and denigrate Africa even further.

These challenges are not exhaustible. So, Africa must address its digital obstacles to become a global player in the tech industry.

References

Abdella, A. (2020, September 2). *How investing in digital infrastructure can make the difference to Africa's economic recovery*. Microsoft. Retrieved May 30, 2022, from https://www.microsoft.com/africa/4afrika/digitalinfrastructure.aspx

AF Bureau. (2021, March 22). *What is a tech ecosystem, and how it can be highly impactful for startups*. ALCOR FUND. Retrieved May 11, 2022, from https://alcorfund.com/insight/what-is-a-tech-ecosystem-and-how-it-can-be-highly-impactful-for-startups/

Ahmed, A. (2020, September 22). *The journey to a peaceful, prosperous and integrated Africa, according to Ethiopia's PM.* World Economic Forum. Retrieved May 11, 2022, from https://www.weforum.org/agenda/2020/09/africa-peace-prosperity-potential-economics-independant

Awanis, A., Lowe, C., Andersson-Manjang, S. K., & Lindsey, D. (2021). *The state of the industry report on mobile money 2022*. GSMA. Retrieved May 30, 2022, from https://www.gsma.com/sotir/wp-content/uploads/2022/03/GSMA_State_of_the_Industry_2022_English.pdf

Bayuo, B., & Bamford, R. (2021, November 3). *What would it take to make Africa a Tech-Startup superpower? Towards a roadmap for policymakers*. Institute for Global Change. Retrieved May 11, 2022, from https://institute.global/

policy/what-would-it-take-make-africa-tech-startup-superpower-towards-roadmap-policymakers

Caballero, A., & Bashir, S. (2020, October 20). *Africa needs digital skills across the economy - Not just in tech*. World Economic Forum. Retrieved May 11, 2022, from https://www.weforum.org/agenda/2020/10/africa-needs-digital-skills-across-the-economy-not-just-tech-sector/

Chakravorti, B., & Chaturvedi, R. S. (2019, December 4). *Research: How technology could promote growth in 6 African countries*. Harvard Business Review. Retrieved May 25, 2022, from https://hbr.org/2019/12/research-how-technology-could-promote-growth-in-6-african-countries

Fredua-Kwarteng, E. (2021, February 20). *How universities can contribute to national development*. University World News. Retrieved May 11, 2022, from https://www.universityworldnews.com/post.php?story=20210219085636646

Ghanem, H. (2020, February 7). *Shooting for the moon: An agenda to bridge Africa's digital divide*. Brookings. https://www.brookings.edu/blog/africa-in-focus/2020/02/07/shooting-for-the-moon-an-agenda-to-bridge-africas-digital-divide/

International Finance Corporation, IFC. (2019). *Digital skills in Sub-Saharan Africa: Spotlight on Ghana*. World Bank Group. https://www.ifc.org/wps/wcm/connect/ed6362b3-aa34-42ac-ae9f-c739904951b1/Digital+Skills_Final_WEB_5-7-19.pdf?MOD=AJPERES

International Telecommunication Union, ITU. (2020). *Measuring digital development: Facts and figures 2020*. International Telecommunication Union Publications. https://www.itu.int/en/ITU-D/Statistics/Documents/facts/FactsFigures2020.pdf

Kyobutungi, C. (2021, May 18). *African countries must muscle up their support and fill massive R&D gap*. The Conversation. https://theconversation.com/african-countries-must-muscle-up-their-support-and-fill-massive-randd-gap-161024

Levi-Strauss, C. (1968). *The savage mind*. University of Chicago Press.

Midega, J., Kyobutungi, C., Okiro, E., Okumu, F., Aniebo, I., & Erondu, N. (2021, May 20). *Countries should up support for research and development*. University World News. Retrieved May 11, 2022, from https://www.universityworldnews.com/post.php?story=20210519073727854

Pazarbasioglu, C., Mora, A. G., Uttamchandani, M., Natarajan, H., Feyen, E. & Saal, M. (2020). *Digital Financial Services*. World Bank.

Rocca, C. (2019, July 11). *Africa's Youth: Action needed now to support the continent's greatest asset*. Mo Ibrahim Foundation. Retrieved May 30, 2022, from https://mo.ibrahim.foundation/sites/default/files/2020-08/

international-youth-day-research-brief.pdf

Signé, L. (2018, December 12). *Africa's consumer market potential*. Brookings. https://www.brookings.edu/research/africas-consumer-market-potential/

STEMpedia. (2019, December 2). STEM education in Africa - the Past, Present, and Future. STEMpedia. Retrieved May 11, 2022, from https://thestempedia.com/blog/stem-education-in-africa-the-past-present-and-future/

Wadhwa, D. (2018, September 19). *The number of extremely poor people continues to rise in Sub-Saharan Africa*. World Bank Blogs. https://blogs.worldbank.org/opendata/number-extremely-poor-people-continues-rise-sub-saharan-africa

World Bank. (2021, January 21). *Tackling Africa's skills gap to build more robust and diversified economies*. World Bank Group. Retrieved May 11, 2022, from https://www.worldbank.org/en/region/afr/publication/tackling-africas-skills-gap-to-build-more-robust-and-diversified-economies

Zeufack, A. G. (2019, June 4). *Taking the economic pulse of Africa* [PowerPoint Presentation]. World Bank Group, JETRO-IDE, Tokyo, Japan. Retrieved May 25, 2022, from https://thedocs.worldbank.org/en/doc/356931559800758619-0090022019/render/060419AlbertZeufackJetro.pdf

Chapter 12

IT Manpower Acquisition

Many organisations and the governments of many African countries today realise that information technology systems, computers and telecommunications products have potential benefits for them. They can increase efficiency and effectiveness in running operations, strengthening management and administration, improving market performance and conferring competitive advantage. As a result, many organisations and government parastatals have incorporated IT systems into their operations and are constantly updating them. However, the whole bouquet of the benefits of IT systems is only sometimes utilised.

One reason for the incomplete utilisation of the benefits of IT systems or the non-successful implementation of IT in many African countries is the lack of skilled IT personnel (Woherem, 1991a). For instance, while there has been a greater awareness of the capabilities and potential benefits of information technology to organisations in Africa, the level of IT penetration in Africa is slow compared to the developed world due to a lack of adequately trained personnel to design and implement IT systems. There is an urgent need to train various IT personnel in Africa as organisations quickly realise the imperative of having skilled IT manpower for their systems.

Given the absence of sufficient numbers of qualified IT personnel customarily required to plan the IT strategy of organisations, analyse the requirements of a system and design, implement and use a system, many systems do not engender the optimum improvement in efficiency and effectiveness that is, sought. The reason countries in the West are prosperous in the use of IT is that their governments spend a lot on IT education and training, thus, bringing about the availability of more skilled personnel in both their public and private sectors. The governments of various African countries must pay attention to developing computer literacy skills and adopt appropriate measures to promote the effective diffusion and use of IT on the continent.

Skilled IT Manpower: A Major Bottleneck

The low and often inadequate design, implementation and use of IT systems in Africa can be attributed to several problems and inadequacies (Woherem & Rowe, 1990). Firstly, most African countries do not have a coherent and explicit IT strategy. This means that many do not have policies or organisations that examine and regulate IT course content, determine the relevance of various IT curricula to industry and the country, set the standard and duration of training required for each type of IT professional, specify the procedure for accrediting IT training institutions, and regulate software, hardware and development standards. Secondly, many African organisations do not plan their IT systems needs strategically. This results in the haphazard growth of applications that often have no real business or organisational justification and have not been prioritised.

Additionally, many African organisations do not carry out an adequate cost-benefit analysis before introducing IT systems: they rely primarily on their gut feeling and the assurance vendors give them. This leads to systems without economic, organisational, strategic or social justification. In the end, such IT systems are abandoned. Organisations and government parastatals in Africa need to regard their investments in IT systems as capital investments that ought to yield economic, strategic, and organisational or user benefits.

Top management awareness of IT and support is assumed to be one of the main contributors to the success of an IT system. In many African organisations, top managers and corporate executives lack a sound knowledge of IT and its important place in modern businesses and organisations. Organisations have to realise that the commitment and cooperation of their management towards IT systems procurement, design and introduction is vital for the success and continued use of IT systems. Executives who are enthusiastic and supportive of IT systems understand the nature and benefits of IT systems and the skills or training required to design and implement them.

The lack of pertinent infrastructures, such as steady electricity and fixed broadband, is another reason for Africa's slow diffusion and successful use of IT. Computers and IT systems need stable and dependable electricity, but the electricity supply of most African countries is erratic. In addition, there is poor network infrastructure, and telecommunications systems are also weak, hindering the effective utilisation of IT systems for retrieving and communicating data or information.

Also, IT systems suppliers tend to have poor after-sale services. They often provide poor maintenance, training and user support and are more interested in selling their wares than giving good service to the customer. The lack of skilled personnel in most African organisations forces them to depend on the vendors (and other IT firms) for their hardware, software training, support and maintenance. The training provided by the vendors is often on their products, i.e. the technical functionality of the systems they are selling.

In addition to all the above, perhaps the most crucial reason for the failure, underutilisation and inappropriate use of IT systems in many African countries is the shortage of IT professionals. Skilled IT professionals are vital for the success of IT projects. The success of IT is increasingly becoming a people issue, the most critical being skills avail-ability, quality of user staff, software quality, managerial awareness and training.

IT Manpower Requirements

The job titles and skills that should generally be brought together by an organisation intending to develop or use IT systems are vast and varied. Users are the employees and top management who will use the implemented system(s) to solve organisational problems or perform tasks. There are active users (those who directly use the system to perform their functions) and passive users (those who plan or use the work performed by active users).

There are also technical computer personnel, who are technically knowledgeable about computers in the sense of being able to analyse, design, implement and test the hardware or software of computers. They include system designers or analysts who analyse business objectives and determine the organisation's information system (IS) and IT needs and strategy. The analyst first analyses the system's technical requirements (and others) and then designs it using such devices as data flow diagrams, activity decomposition diagrams and entity models. The programmer writes the instructions that the computer carries out. The hardware of a computer will do nothing unless it has been told what to do by a programme written by a programmer.

The IS/IT manager selects and directs a group of staff to work collectively to attain set goals within previously determined budgets and timescales. This means that the manager has to be acquainted with techniques for costing, estimating and monitoring money and timescales and for managing people. It is often tough to discover a manager knowledgeable in the technical, business, and managerial aspects of IT. Today, there is a consensus that the IS/IT manager should have more knowledge of the latter than the former. The project manager is the one with the overall responsibility for the success of an IT project, defining objectives, producing the technical and quality plan, monitoring progress and appointing other staff members.

Electronic engineers are responsible for piecing together the different components of IT equipment. They diagnose faults and perform routine maintenance on physical IT equipment. Electrical engineers determine the necessary power supplies for a system and the level of loadings on computer room floors.

Other personnel who play or should play a role in an organisation's IT systems development are human factor professionals who are well-versed in the ergonomic, social and organisational aspects of IT systems design and, therefore, ensure that IT systems are designed to be user and organisation-friendly. Legal personnel are needed to determine the contractual conditions for a project or computer equipment purchase and to interpret contract conditions about their impact on the organisation or project. Administrators supervise the project or operation procedures to perform tasks efficiently and expeditiously. Accountants monitor and analyse the financial accounts of a project or organisation and determine the financial means of meeting objectives.

As it has already been pointed out, most African countries have a shortage of skilled personnel in these areas of IT. Even more so, they lack people at the top computing strata, such as IS/IT strategy planners, managers, project managers or leaders, and human factor professionals. The universities, polytechnics and other IT training institutions in most African countries provide ill-trained and under-skilled professionals; hence, the systems they develop are often of poor quality and underutilised or ill-used. Looking at even the best training institutions in most African countries shows that they provide only a rudimentary level of IT training and education compared to their equivalents in Western nations.

The IT and computer science training curricula in universities and polytechnics usually lack courses on the organisational and social aspects of computing. In other words, the training students receive does not include how to have non-functional issues in the requirements analysis, specification, design and use of IT systems. IT and computer science students may take a business course that is not taught in alignment with IT. It is not a business computing course and does not include knowledge of how to carry out business analysis and IT systems cost-benefit analysis. In fact, most of the courses offered by IT training schools and universities in Africa tend to be too technically biased.

These institutions have limited scope for understanding the organisational, business cum economic and social aspects of IT. Thus the graduates of these institutions often find it challenging to understand the social implications of the technology and how to tailor it creatively to solve problems specific to their organisational, social and cultural environment. There is a growing understanding in the West that while prior technical considerations inform most IT systems introductions, they fail because of inadequate attention to organisational, business and social factors (Woherem, 1991b). It is now advised that these non-functional issues should be considered in the requirements specification of systems. The IT curriculum used by universities and polytechnics in Africa should include components on these non-technical aspects of IT.

A Strategy for IT Manpower Acquisition

Given that the successful use of IT systems in an organisation depends on the availability of enough skilled personnel to plan, analyse, design, implement and use the system, organisations should plan their manpower needs to reflect their IS/IT strategies. These, in turn, should be based on the business strategy (see **Figure 28**). As there is a shortage of IT professionals in most African countries, organisations should ensure that they have a recruitment strategy for their IT staff and that their staff are motivated with good salaries and incentives. Organisations will gain the most from IT when the design and implementation of IT systems and the manpower requirements for the task have been part of a carefully thought-out overall strategy through a thorough planning and consultation process.

Organisations should not invest in the training of IT personnel on the rationale that a competitor is making an investment in IT, recruiting or training IT staff, or as a result of the technical dazzle of an IT systems demonstration, or as a result of the sales pitch of IT systems vendors. They should not let a feeling of being left behind dictate their direction in IT or the types and number of IT professionals to recruit. Instead, the approach should be based on deliberate and carefully drawn-out personnel plans to meet the organisations' needs. It is essential to relate IT to the organisation's business goals; the question to be asked is: how can IT help to achieve them? How can IT help carry out the organisation's basic activities to meet the objectives?

Formal IT systems can trigger the need for IT manpower planning, new projects, changes in the organisation's functional units or business process, external threats, requirements, thrust or legislation. For example, changes in the organisation structure may trigger changes in the resource availability for software development or product support, and so changes in the manpower requirements.

The IT Strategy Planning Process

As the IT manpower strategy of organisations needs to be tied to both their information systems and IT needs, the starting point of the IT strategy planning process should be the business needs and strategy of the organisation. In other words, business development strategy has to define the technology and manpower imperatives. The aim of starting the process of IT systems acquisition and personnel planning from a company's business objectives is to align the business needs and the IT systems of the organisation, to support the attainment of the organisation's business objectives (see **Figure 28**).

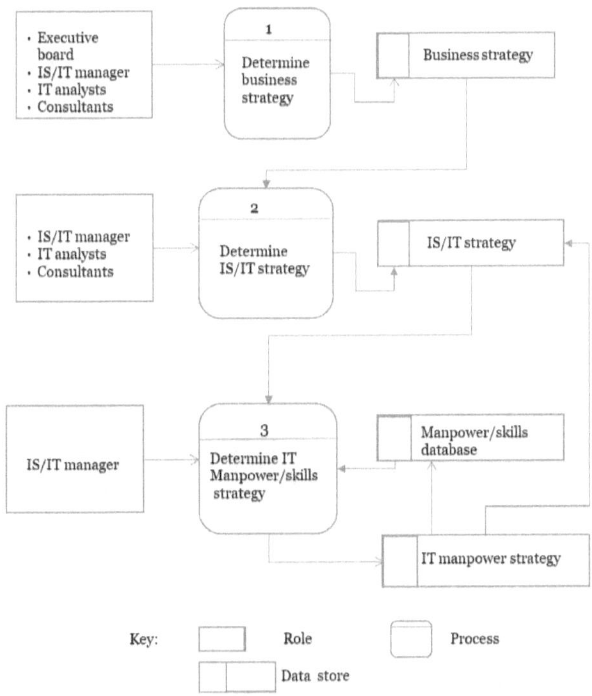

Figure 28: Strategy for IT manpower needs acquisition in an organisation

The development of an IT manpower strategy begins with the definition of business objectives and strategy for the foreseeable future, for example, five years ahead. The business objectives take into consideration market trends, the competitive position of the organisation, moves of other organisations in the market, technological developments, and financial capability of the organisation, infrastructure, policy and standards, the state of growth or maturity of the organisation, its competitors, and any other environmental factors which are likely to affect its future. The whole process of aligning the IT systems needs of an organisation to the business needs is usually referred to as IT strategy planning. Generally, an IT strategy of an organisation is a blueprint for developing an information technology capability to support the organisation's business or corporate goals to create sustainability and maximise value return on investment.

IT Manpower Strategy Planning

IT Manpower strategy, a part of IT strategy planning, specifies how to recruit and place IT personnel in an organisation. After determining the IT strategy, the question is: do we have the manpower and skills needed to implement the set of applications suggested by our IT strategy over the planned period? Given that the answer is no, the next question is: how do we ensure that we do?

Once the IT needs and strategy are known, the IT manpower available to an organisation has been determined, then the IT manpower/skills requirement estimate for the given plan period can be determined. The public resource profile of the organisation is examined to see how it can match the estimated needed manpower and skills. The shortfall is then ascertained, and a plan for satisfying it is worked out (see **Figure 29**). The report detailing the shortfall and a plan to close the gap should contain information on what has to be done, who should do it, the deliverables, dependencies and estimates of the time and effort required. It should also stipulate task definitions and comments on why specific personnel are suitable for some identified tasks.

The information concerning manpower and skills availability should be stored to reflect the organisation's IT development and support infrastructure and types of projects or activities. For example, it can be organised according to groups, teams, technical support, projects, customer support, educational qualification, management, quality assurance, programming skills, language skills, analysis skills, design skills, years of experience, support or use of applications and database skills. The manager should select the team for the project, allocate people according to their expertise and skill, construct a detailed project schedule, and shuffle the work to reduce gaps and optimise timescales.

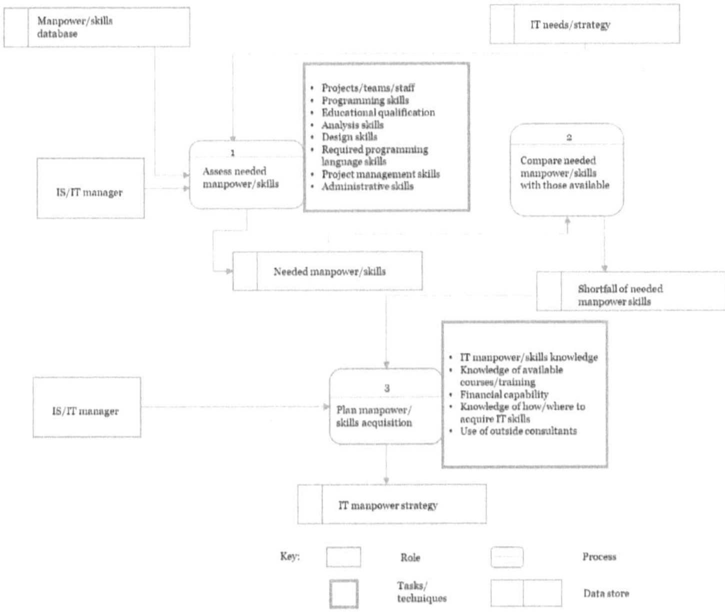

Figure 29: IT manpower needs planning

To ensure the availability of their IT manpower and skills requirements, organisations can make use of the following instruments or policy options:

- Give employees internal courses or training for specific job profiles.
- Give employees external courses or training.
- Hire new job applicants. Pertinent experience, qualifications and skills requirements that will help fill the jobs should be defined.
- Take over the basic training for some jobs. This could be linked with courses in external educational institutions. Personnel could be placed in other organisations or training centres.
- Some personnel could attend mid-career, full or part-time courses which provide academic or practical knowledge or training.
- Employ some personnel to "learn by doing." Some inexperienced staff can be made to participate in projects with the supervision of more experienced personnel and, by doing so, become skilled in the area. They can also get more formal training or education in-house or externally.

- Employ the services of external IT consultants.

Generally, it is advisable to train existing staff rather than to recruit new ones. This is a better policy because the existing team already knows the job and is committed to the organisation. It can be expensive to recruit new employees who need ample time to get acquainted with and adjust to the organisation. The main attraction of internal training is that the trainees acquire skills relevant to the organisation's specific needs and the general abilities obtained from educational institutions or establishments.

Staff members such as secretaries, clerks and administrative workers can be retrained to acquire IT skills. To ensure prudence in managing financial resources, it is wise to ascertain whether organisations can satisfy their IT needs from internal retraining before venturing into the IT Labour market.

The internal or external training of personnel has its drawbacks in terms of the economic justification of such activity to the organisation in the long term. People with IT skills can quickly move from one firm to another. There is a strong tendency for IT personnel to leave the organisation that trained them to complete their training and exploit their newly-acquired skills in another organisation offering better salaries and working conditions. This is less than satisfactory for the organisation that trained such personnel since the money and time invested in training will have been wasted. The most crucial route to retaining trained workers is to offer salary levels and working conditions above the norm.

External hiring of skilled personnel can be acquired by placing an advertisement, poaching from other organisations, or hiring a head-hunting firm to seek suitable employees.

The services of an external consultant can be employed when an organisation cannot provide the internal staff to carry out the task satisfactorily, or the internal team is not available for the project because of other engagements; or the organisation wants to gain an objective, external view. To utilise consultants effectively, their tasks should be well-defined. They should be given information about the company's products and working methods. They should interview relevant staff and ascertain as much background material as possible to give their opinion on the project's feasibility and necessity. They should be employed under terms of reference. These should include the duties of training a permanent staff member who will understudy them.

Furthermore, they should be provided with details of the organisation and its structure, sufficient to clearly understand the organisation's business and the nature of the task. Finally, they should be chosen based on their experience with similar jobs, their knowledge of hardware and software, if already chosen, and their financial stability. It should be remembered that good consultants will undoubtedly be able to display a portfolio of satisfied clients.

The use of external specialists on short-term contracts should also be considered. Contractors tend to be knowledgeable and competent. They should be seen as temporary staff. However, despite that they are ideal for creating and developing a project within a given time, they could be more satisfactory for long-term employment. Contract staff can only be satisfactorily employed where there is a substantial amount of work, and each specialist can be deployed to deal with a specific aspect of the project. Project tasks should first be divided into attainable modules before recruiting contractors.

Macro-Level IT Policy Planning

Governments of African countries have a role to play in ensuring that IT gets well-diffused, developed and used in their respective countries to enhance their national development objectives. However, organisations (large and small), as an interest group, have a role to play in assisting the governments of their respective countries in determining their IT manpower and skills requirements. By synergising and articulating their present and future IT needs, they can help governments assess overall IT manpower and skills needs, determine a suitable curriculum for satisfying them, select the accreditation procedure for IT training and educational institutions and decide the number of candidates to be trained in a year. By so doing, organisations will enable their governments to determine and provide the type of systems they need and the amount and level of necessary infrastructure for the systems to operate. The government has to understand what is required by the IT community and the importance of satisfying them, for it is only then that it can weigh those needs with other divergent demands of the economy, given its financial position, and arrive at a plan, financial budget and action to satisfy those needs.

Organisations should also help their government provide, where possible, some of the finances needed and the planning expertise required. The governments should establish some IT excellence centres in the cities and regions of the country. Governments should have a representative of the IT-user organisations on their national and regional IT committees. The committees should plan IT infrastructure, including electricity, education and training, and telecommunications facilities.

National and regional IT centres should have resources to provide training for IT trainers, users and technical IT personnel. Organisations can influence the profile, content and quality of courses offered by institutions and the views and actions of the national and regional IT committees by being members of such committees.

Finally, the government should be encouraged to improve the salaries of IT personnel in their employment and other incentives for them to work in the industry. They should also enable organisations to follow suit so that skilled IT personnel are not lost to other professions or sectors.

References

Woherem, E. E. & Rowe, R. (1990). *The problem of developing workable IT systems in Africa*. Paper presented at the Nigeria Institute of Marine Engineers/ Lloyd's Register Seminar, October 24th, Lagos, Nigeria.

Woherem, E. E. (1991a). Human factors in information technology: The socio-organisational aspects of expert systems design. *AI & Society*, 5, 18-33. Retrieved September 1, 2022, from https://link.springer.com/article/10.1007/BF01891355

Woherem, E. E. (1991b). Information technology and Africa: An appraisal of the present situation and future potential. *Project Appraisal*, 6(1), 33-45. Retrieved September 1, 2022, from https://www.tandfonline.com/doi/abs/10.1080/02688867.1991.9726790

Chapter 13

Procurement and Design of IT Systems

Information technology procurement is the process or series of activities involved in sourcing and acquiring IT assets to drive an organisation's operations. Typically, IT assets and services are categorised into hardware and software, which include, but are not limited to, the following:

- Desktop computers and laptops
- Peripheral computer equipment such as keyboards, printers, copiers, cables, and adapters.
- Software development or procurement services
- Switches, routers, controllers, etc
- Data storage
- Internet Service Providers (ISP)
- Telecommunications equipment and services
- Computer data input services
- Audio visual equipment and services
- Web services
- IT Consulting
- Maintenance contracts
- Network devices
- Cloud services
- World Wide Web site operation (host)

IT procurement aims to fulfil an organisation's IT needs to carry out business smoothly. Usually, IT procurement is aligned with IT asset management (ITAM) to ensure that, firstly, an organisation gets maximum value out of its spending on its IT assets. Secondly, ITAM protects an organisation's assets from regulatory and cyber security risks.

Stages of an IT Asset Lifecycle

Every IT asset has a life-cycle, a period within which it can be put to optimum use. While each organisation may have unique life-cycle stages for their assets, the life-cycle of an IT asset generally includes the following stages:

1. **Planning** – This has to do with figuring out the assets needed by the organisation, what they are required for and how they would be acquired.
2. **Procurement** – This is the series of activities that are undertaken to acquire the IT assets needed by an organisation. IT assets acquisition could be via building, buying – including software-as-a-service, licensing, or leasing.
3. **Deployment** – This refers to putting an asset to the use for which it was acquired. Asset deployment may include installation, interaction with other technologies, or authorising user access.
4. **Maintenance** – After asset deployment, provision has to be made for maintenance, which may involve repairs or upgrade to maximise its usage and life-cycle.
5. **Retirement** – Typically, an asset offers maximum value over a duration. At a certain point, depreciation will set in so that acquiring a new asset helps to make more business sense than an insistence on continually repairing the old one. An organisation may retire an asset because a better version has been developed. On other occasions, creating a new version of an asset may not require the old version to be retired; it would only need to be updated.

The IT Procurement Process and How to Manage It

Every organisation has to have an IT procurement process to acquire necessary IT assets. This process must be managed responsibly by seeking suitable suppliers to ensure that the needed items are obtained at the best possible prices (see **Figure 30**).

Generally, organisations adopt the following process when procuring IT assets:

Step 1: Identifying Need – This involves collaborating with internal teams to identify the IT product or service needed.

Step 2: Review and Approval – After the internal teams have put forward the IT product or service needed, a review process is undertaken to justify the need. The approval process involves using techniques such as cost-benefit analysis to evaluate the benefits and risks of procuring the identified product or service, after which approval is given for an investment.

Step 3: Selecting Suppliers – This stage is where suppliers are identified and contacted about supplying the product or service. The suppliers selected could be new or existing ones with whom the organisation has a relationship.

Step 4: Negotiating – The selected suppliers are given sufficient information to allow each of them forward a realistic quote or bid. The quotes are evaluated, and successful bidders are selected, after which terms are negotiated to get the best possible deal. Once an agreement has been reached, an IT procurement contract is signed for the product or service to be delivered.

Step 5: Placing the Purchase Order (PO) – This document defines the price, specification, and all other terms and conditions of the product or service to be supplied. All obligations to it are included in the PO.

Step 6: Receive and Integrate – The PO is conveyed to the vendor, who will acknowledge receipt and express commitment to deliver the product or service as contained in the PO.

Step 7: Expediting the Purchase Order – Sometimes, it is necessary to expedite the purchase order, for instance, if a product that has been in use deteriorates faster than anticipated. In such an instance, the timelines could be examined to adjust the delivery and payment dates.

Step 8: Receipt and Inspection – Once the product has been delivered, a review is carried out based on the specifications outlined in the purchase order. The invoice and receiving document are examined against the PO to ensure no discrepancy between what was specified in the PO and what has been delivered. If there is a discrepancy, a receiving discrepancy report may need to be filed accordingly.

Step 9: Invoice Clearing and Payment – If the assets supplied match what was specified in the PO, the organisation would accept the product or service, which means they will be obligated to pay for it. Accounts payable should be aligned to complete the buying process.

Step 10: Maintaining Records and Relationships – Care is taken to keep proper records for the audit, tax payment, and product warranty confirmation. Going forward, most organisations endeavour to maintain a cordial work relationship with the suppliers for further business transactions.

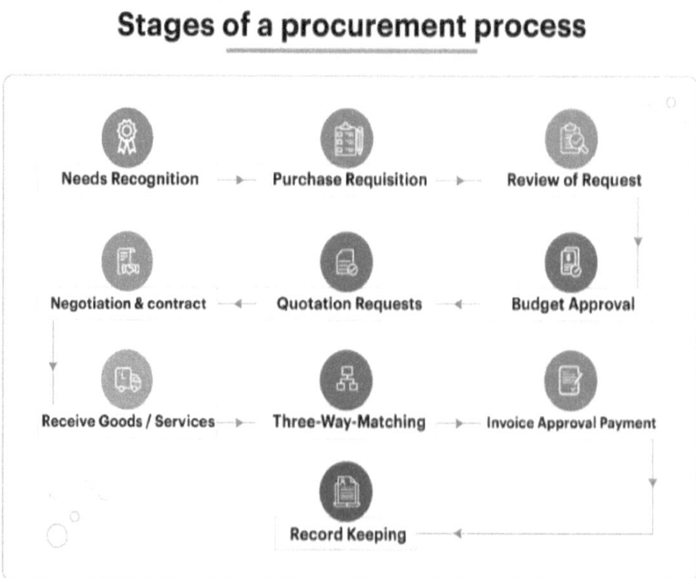

Figure 30: Stages of a procurement process

IT Procurement Best Practices

The following are a few recommended best practices for IT procurement:

1. Acquire Technology That Accurately Fits the Organisational Needs and IT Environment – It should always be the case that the technology acquired is strategically procured to fit and support the organisation's goals. Procuring new IT assets only to have an engineer modify the current IT environment to accommodate the new asset can lead to inefficient output and integration problems.

2. Form Strategic Relationships With Vendors – It is essential to forge close relationships with suppliers and IT providers. The numerous benefits of such relationships include cost savings by leveraging economies of scale, more efficient resolution of adoption problems, and establishing service level standards and agreements.

3. Engage Stakeholders in Procurement Decisions – It is crucial to garner as much data as possible from all relevant stakeholders when making IT procurement decisions. By involving as many stakeholders as possible, more people involved in the procurement process will have a fuller knowledge of the organisation's budget needs and how the assets to be procured will fit the IT environment of the organisation.

4. Mitigating Risk – No matter how thorough the procurement process is, there is always an element of risk attached to IT assets acquisition. It might be that the technology acquired may fail to fit in adequately with the organisation's needs, or some compliance issues may arise in the course of using the IT asset. Identifying potential risks and addressing them ahead of time is a critical IT procurement best practice. This is done by conducting a complete risk analysis, deciding potential financial losses and creating and implementing a risk mitigation plan.

5. Acquire Assets with the Long-Term Future in View – Organisations should seek to implement a procurement system that acquires IT assets in the short and, most importantly, for the long term. Acquisition decisions based on short-term goals may necessitate some upfront expenses and lead to a reduction in capital acquisition costs. However, it is more prudent to consider the total cost of ownership when acquiring IT goods and services.

Acquisition of New Software

Organisations must acquire new IT systems to replace their old manual systems when they have ascertained a need for such a new system. In this case, the organisation would have to purchase the new system from the marketplace, have consultants and independent systems developers, or develop it in-house. Even when existing IT systems are used and constantly updated to meet the company's changing needs, there comes the point when much consideration should be given to obtaining new software, for example. If the existing system is inadequate to meet the requirements or if the current software is so outdated that the organisation's sustainability is compromised, the procurement of new software should be considered.

Managers and software developers are primarily interested in determining whether or not a software package will meet all of the criteria. In most situations, the software package only solves a subset of the requirements. As a result, either the acquired software must be updated, or additional software must be installed. Like their counterparts in other parts, African organisations are developing software at an increasing rate. Organisations procuring new systems need to answer some of these questions before the procurement:

- How well can we use the software as it is?
- How superficial is it to keep the software up to date?
- How well will we be able to use it?
- Will we be able to use it even if the processing environment changes?
- Is it possible to make changes to the purchase or lease agreement?
- If not, are there sufficient indications that the package's vendor will make the necessary changes?
- Will training be required for the personnel to use, adapt, or verify the software package?
- If so, would the training be offered for free or at a cost to the buyer?

After some time, some software becomes error-prone, ineffective, and costly to maintain. The costs and benefits of the continued maintenance of such software need to be appraised and used to decide whether to purchase a new system or continue using the old one. Although there are no particular rules for determining when to rebuild rather than maintain existing systems, organisations should decide to change to new software if one or more of the following conditions applies to their existing software in usage:

- repeated failures
- programme that is up to ten years old
- too complicated software architecture and concepts
- code developed for outdated device
- operating in emulation format

- massive configurations or subroutines
- Excessive resource constraints
- difficulties in retaining developers
- hard-coded specifications that must be altered
- inadequate documentation
- a poorly communicated design specification

A practical application system has an estimated life-cycle of approximately seven to ten years, while 15 to 20-year-old software applications are not uncommon. Because of the frequent changes and upgrades, software tends to degrade with age. However, if the software was designed and built methodically and sustainably, and if maintenance was diligently conducted and recorded following established standards and guidelines, it may be conceivable to keep it running effectively and efficiently for several more years.

Towards Effective IT Systems Design

IT systems design establishes, integrates, and implements computer systems to meet an organisation's specific goals and specifications. It entails specifying and designing system aspects like components, modules, interfaces, data for a system, and architecture.

As highlighted throughout this book, information technology has become an integral part of our lives in the 21st century. It is central to the world's economy and is mainly responsible for making it a global village. For African countries to fully integrate into and benefit from the global economy, their scientists and engineers must become adept at designing IT systems that meet the needs and requirements of both local and international organisations.

The design of information technology systems has been informed by considering functional (task performance) issues to the detriment of non-functional (social, organisational, business and ergonomic) issues. In the process, IT system designers realise that systems fail when not designed to satisfy the needs of organisations and users. As a result of increasing awareness of the folly of a techno-centred design paradigm for IT systems, some new software systems development methodologies were developed with the promise that they would help ensure that not only technical functionalities are analysed and specified but also that socio-organisational requirements are considered and specified in the development of systems.

Technical, economic, organisational and human design goals or considerations should all be balanced when designing IT systems for optimum efficiency and benefit to the organisation and the user. In other words, there is a need to balance the goals of the organisation and those of the users, such as users' need for skilful, complex and flexible jobs where they can exercise job control, initiative and judgement. This balance is not only possible but mandatory.

As a result of the technical bias and the non-consideration of socio-organisational issues in the design of IT systems, many systems fail to be accepted in the long term, either by the management of organisations or by the users. Some projects are stopped halfway without being completed due to ignoring the socio-organisational requirements of the systems. Many systems fail to be accepted by management as they are deemed deleterious to the business and managerial needs of the organisation. Also, many systems fail to be accepted by the users because they are perceived to threaten their jobs, skills, job control and job satisfaction. Some systems are deskilling and cause job dissatisfaction for the end users (Woherem, 1988).

Many systems are not aligned with the business objectives and critical success factors of the organisations they are being used in. While many systems are technical successes, they are failures at the organisational, social, economic, or business levels (Woherem, 1991c). There are cases where an IT system is introduced when a mere restructuring or reorganisation of the enterprise would have been preferable, a conventional system would have sufficed, or a different system could have been better in meeting the organisation's objectives. Many managers are unhappy with the systems they are saddled with in terms of performance and physical contribution to the productivity and strategic objectives of the organisation (Thomas, 1991).There is often much disillusionment with IT systems and their rampant failures, which are blamed on the design of the systems, when in fact, the failures result from the myopic and technocratic paradigm employed in their design.

Socio-Organisational Aspects of System Design

There are socio-organisational aspects that need to be considered during systems design, development and introduction (Woherem, 1991c). A socio-organisational orientated systems design emphasises these issues in addition to the traditional technical functionality issues.

Organisational Requirements Definition for Information Technology (ORDIT) was a five-year project in Europe. It sought to assist IT systems designers in understanding and capturing all the necessary socio-organisational factors needed in systems design and the notations for analysing and specifying them. The project started in January 1989 with academic and industry participants. These included the HUSAT Research Institute, MARI Advanced Technologies Limited, the Computing Laboratory of the University of Newcastle UponTyne, the Work Research Centre Limited, and Algotech Sr1 (Poulson, 1990).

Social issues concern the system's impact on the skills, job control, job satisfaction, job performance and ownership of the users. These issues affect users in a manner which determines whether they enjoy their jobs or not, whether they are alienated from their jobs or not, or whether their jobs become more or less constraining or deskilling as a result of the machines used in their domain.

Organisational issues concern the manner that affects the structure, managerial strategy and philosophy, group dynamics and culture of the organisation. They determine whether the system will restructure the organisation, impinge on its executive style, change its culture or affect group cohesion and dynamics. Most methods involve all these factors, but the question is how a system affects them, for ill or good.

Business issues include the financial capability of the organisation, its profitability, business objectives and strategy and the return on investments in the system. These issues are also significant for the success of an IT system in an organisation. What is the use of introducing a plan that will not bring any financial, strategic or business benefit? IT systems must be well aligned with the business strategy of an organisation. They should be introduced in collaboration with the business objectives and design of the organisation.

Technical issues include accurately modelling and representing all the technical functionalities of the system in the systems requirements and design documents or files. It includes the technical requirements, the system's user interface, and the organisational requirements.

These four issues affecting the design and use of IT systems in organisations are interrelated and mutually adjusting. They act and are affected by each other.

Satisfaction with these issues is difficult, as they make differing and conflicting demands. What might be technically correct and helpful for the system, or enlightening or challenging for the designers, might be socially deskilling or job constraining for the users, or deleterious for the organisation. To achieve optimum satisfaction, conflicting needs should be balanced and quantified. Technical issues are always considered. Some business issues are sometimes considered before the design of systems. However, social and organisational issues affect the long-term acceptance of systems. Therefore, all these issues must be considered, weighed, prioritised and used in the systems' requirements analysis and functional specification before their introduction. Designs that incorporate holistic requirements, as described above, are beneficial and lead to systems that ensure technical, organisational, business, and user successes.

A Holistic Methodology for Systems Development

The issues and stages involved in the design of systems suggested in the diagram below (see **Figure 31**) are not sacrosanct; they can be tailored to suit the needs of each organisation. They are primarily proposed for the development of large systems for large organisations. It might be difficult for smaller organisations to spend the time and resources required to design a system using the prescribed development methodology. As a result, the approach provides a set of concepts and procedures that may be implemented in many organisation settings and tailored to meet specific needs.

The methodology specifies how systems should be developed, the stages involved, and what to do in each step but does not describe how to carry out each task or the precise notation to perform the tasks. It refers to other literature where necessary and leaves it to the designers and their organisations to determine how to perform the functions at each stage. This is intended to enable the methodology to be simple, flexible and tailorable. Also, it does not explore the pitfalls of carrying out each task, nor does it say what not to do at each stage. It is simply a general design framework or sketch of necessary development stages in the IT systems life-cycle and what to do in each.

The methodology is based on the perception that systems development is a hybrid process involving IT professionals, users, and the organisation. There is a need, therefore, for the joint optimisation of the interdependent social, organisational, business, and technical subsystems. Thus, it incorporates all these aspects of systems design and development so that the systems developed through the methodology can satisfy the organisation's actual needs.

It has as its conceptual framework the influence of concepts and ideas from numerous fields and studies of methods, techniques and approaches used, or proposed for use, in software systems and other technical systems design literature. These include systems approach, planning approaches or strategies for IT systems planning and management, participative design by users, prototyping, structured techniques, analysis, design and development, and open systems (i.e. the idea that organisations are open-ended systems affected by external factors), purposive systems, and labour process theories (Rosenbrock, 1987, 1990).

Particular precepts and propositions also inform this methodology for good socio-technical systems design. When developing administrative systems, socio-technical systems design (STSD) techniques consider personal, societal, and organisational factors as well as technological factors (Baxter & Sommerville, 2011).

Adopting a socio-technical approach to system development results in more appealing solutions for target consumers and significant stakeholder benefits. The core concept of the socio-technical methodology is that systems design should be a technique that considers both social and technological factors that influence the creation and use of technical systems. Socio-technical approaches to systems design are motivated by the assumption that IT systems may fail to assist the organisation in achieving its objectives if they are not used or useful. Most IT systems satisfy their technical criteria, yet they are regarded as failures because they do not support the actual operations of the organisation.

Users often reject a system that is difficult to use or devalues their work. Thus, the effective use of such technologies depends on deploying a planned change strategy that fits the demands of people dealing with significant changes in their professional careers, as the ability and desire of personnel to use modern software to engage in productive tasks are required for effective IT maximisation. The participation of all relevant stakeholders in the system's design is essential for developing effective socio-technical systems.

People involved in design processes must progressively adopt the concept of designing socio-technical systems instead of developing technical systems. By so doing, technical personnel preserve existing skills and techniques while investigating how those correlate to new or extra strategies required to cope with human and organisational challenges. Some methodologies are designed to be unique and different. This means that designers must give up their existing technical-biased methodologies for more holistic methods. Despite the above best practice, what usually happens is that developers select what they think is practicable and valuable in any methodology, incorporate them into their current systems development repertoire, and jettison the rest.

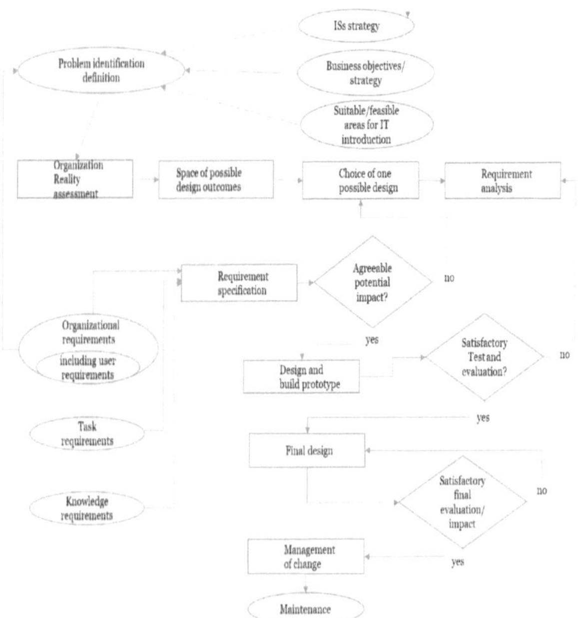

Figure 31: Socio-organisational design of IT systems

A Holistic Methodology for System Development

This methodology recognises the open-ended nature of modern organisations, the fuzziness of the factors that affect or are affecting them, and their interrelationships and counter-effects. It is nevertheless prescribed within a structured methodology so that systems can be developed within the knowledge of IT professionals or the methodological frameworks of systems developers and because most managers prefer the use of structured methods in their systems development projects. This is because most managers perceive structured development methodologies as more capable of ensuring that projects are carried out within planned costs, quality and deadlines; they have more visibility of the stages and milestones of the project, and they maintain better control of the processes of the project.

Users and other stakeholders should be consulted at each stage of the development life-cycle: during the identification and definition of the problem, the choice of solution, the specification of the requirements and functionality of the system, the design of the system, and the specification of the type of user interface to be employed and the environment in which they are to be used. Also, it is essential not to discourage the expert, the managers, or the users, in discussions at each stage.

The first stage in developing an IT system should be identifying and defining the problem. This stage involves understanding the mission of the organisation, the business objective of the particular strategic business unit (SBU), the external factors that affect the product or service and the business success of the SBU, the critical success factors (CSFs), and the strengths, weaknesses, opportunities, and threats (SWOT) of the SBU or department of the organisation.

The strengths and weaknesses of the organisation should be used to match the opportunities and threats posed by the business environment. The value chains (i.e. all the activities performed to design, produce, market, deliver, and support a product or service), SWOT and CSFs then determine the activity areas that are most important for achieving the business objectives of the organisation.

The processes, objects or entities and their relationships within those activity areas and the nature of the information flow in them are ascertained, thus becoming the information system (IS) requirements of the organisation, department or SBU. An analysis of the CSFs can show whether the types of activity requirements in the critical area require IT for their solution. This can be ascertained by asking whether the area is suitable and feasible for introducing IT. In that case, whether the system can be organisationally, socially and economically viable or justifiable in that domain.

Some of the rules for making such a decision can be used by human decision-makers in this area to ascertain whether the system will be viable before more concrete decisions are made towards the development of the system, which might be too expensive to go back on. It is better to discover potential issues before developing an IT system. After the analysis of the business objectives and strategy of the organisation, the study of business areas that require IT solutions is carried out. After this, for each business that needs an IT solution, the analysis of each problem the new system will handle is carried out to ascertain its suitability and feasibility for the particular solution to be used. The designers are satisfied that such a system will either be of business, social or organisational benefit. An in-depth assessment of each issue is conducted to ascertain the organisational reality of the proposed system.

To determine the organisational reality of a proposed system, one must find all the competing forces or requirements for the system to be introduced. These forces are the technical issues involved in the design of the system, the organisational forces, the social forces, and the business issues involved in developing the particular system. All the benefits to each of these areas are determined; the fears, doubts, requirements and demands are also resolved and considered. This determines the factors to be satisfied in the later stages of the design of a system.

Different design possibilities or scenarios can be chosen, and their consequences to each of the above issues are determined. This will lead to optimal or best design choices for the organisation and engender the best solution. An analysis of the documents produced from the assessment of organisational reality will show that these factors make conflicting demands. There is no easy way of satisfying one area without compromising another. This situation will bring about possible solutions, showing how the concerns of the four issues – social, organisational, business, and technical – can be satisfied. One of the possible scenarios is then chosen, leading to actual requirements analysis for the domain area.

The requirements analysis leads to the development of requirements and functional specification documents for the system. The document is similar to that produced during conventional systems development. It is usually used as the base document for the system's design. It includes components of the problem identification and definition stage – the feasibility, suitability and viability of the system to be developed. It also includes:

- A description of who the end users are.
- What the needs of the users are.
- What the needs of the organisation are.
- The nature of the user interface. This calls for the indication of, for example, the required level of computer experience of the users, the interfaces to any systems they currently use and the screen layout and help/explanation facilities that are needed.
- What the response time should be? In other words, what will the users accept as the optimum delay time?
- The nature of the environment. The environment and the conditions in the system that will be used are derived from interviewing the users, the managers and the experts.
- The number of people that will use the system.
- The type of peripherals the system should have, for example, whether it should have a QWERTY keyboard or not.
- The required hardware platform. The hardware required for developing and delivering the system (if different) is described in detail.
- The software platform. All the software needed for the development of the system is indicated. It is also necessary to include the implications of maintenance and support costs and the type of delivery system required (if necessary).
- Personnel. It is necessary to indicate the amount of time that management, experts, users and IT professionals will spend during the period of the development of the system.

During this stage, task analysis, knowledge analysis and acquisition (if it is a knowledge-based system), organisational analysis, and modelling techniques are used to ensure that user, administrative and technical requirements are considered and incorporated into the system's design.

System requirements are essential because they enable the designers and users to know precisely the system's functionality and what it should or should not do. The requirements analysis of the system starts with the chosen design outcome guiding what kind of requirements to select for the system. These should include not only the technical needs but also the organisational and social requirements of the system. This means that apart from the traditional technical functions or task modelling and analysis, there should also be an organisational modelling and requirements analysis.

This will help to determine the operational constraints and requirements of the system, such as the organisation's type and structure, as well as how the task area fits within the framework. It will also determine the relationship with authority in the department, how that will affect the system and the implications of this on the system's plan (such as the assessment rules, culture, and values of the organisation). These issues are essential to be considered, as they might affect how the interface, ergonomics and aesthetics of the system will be designed. It will also determine the type of managerial style required to be implemented through the system: whether the managers want the system to store the daily problems and decisions made by the users, who the users are, the times they will work, for how long, and so on.

Social requirements that should also be captured are how the users want to relate to the system – expert to expert, colleague to colleague, student to teacher, or client to consultant. It will show which skills and job control users want to retain and those they want the machine to handle. Also included are task requirements to satisfy the machine users; this is done mainly through task analysis. This will help to derive the task and user requirements of the domain experts or future users to determine how the system will be used, the tasks to be supported by the system, the interface of the system, and so on.

The approach whereby requirements are determined just before starting the design may be based on the fallacious assumption that conditions are fixed. The reality is that it needs to change continually, often in response to organisational change. These changes are more likely to emanate from experience in using the system than from an abstract specification in the early design of the system. The major problem in requirements assessment may not be the development of a complete, consistent and unambiguous specification before design but rather the evolution of requirements which allow a timely response of the software to organisational change. Requirements assessment during maintenance may be as demanding as that during development.

The primary objective of this phase is to provide the information necessary to accomplish a detailed plan. The user (the person who initiated the request) must be consulted. After all of the users' needs and requirements are understood, the following must be conducted:

- Workload estimation and refinement
- Project team staffing and organisation
- Personnel requirements
- Facilities requirements
- System software requirements

After all the requirements are understood, a written plan should be prepared, including all the user requirements and those listed above. This written plan should be included with the documentation.

A preliminary impact analysis of the potential system on the users and the organisation will show whether there is a need to change the requirements or design of the system. It is carried out to ascertain how the system will affect the users, the organisation, the solution to the domain problem, user and organisational performance in terms of productivity, user and corporate acceptance, user job satisfaction, user job skills and job control, the business requirements of the organisation, and so on.

Rapid prototyping is beneficial in systems development. This is primarily related to the nature of human expertise. There are both explicit and non-explicit or tacit dimensions of human expertise, and most experts find it difficult to verbalise their knowledge; even when they understand what they know, it is often challenging. This makes knowledge acquisition or functionalities determination an unreliable and problematic issue in systems development. It is unlikely to be derived correctly from a first-pass requirements elicitation and analysis. It is always advisable to assume that the first set of requirements model is wrong and to go back to the experts or user to ratify what has been modelled.

The prototyping process involves an iterative life-cycle model in which stakeholders and IT professionals agree on the prototype's objectives, scope and criteria for evaluating it. All this information is contained in a written (definition) document and will be updated during each prototyping cycle. The prototype is then constructed after some requirements analysis. This prototype stage includes functional specifications, design work, and testing and fine-tuning tasks. The prototype is then analysed to check its success or failure. The objectives of the definition document, its scope and evaluation criteria are used to evaluate the prototype. Based on this, a decision is reached as to whether to start another cycle of prototyping or not.

Test and evaluation will validate and verify the technical performance of the system. The prototype system enables the experts and users to test and evaluate it so that they can assess whether it accurately reflects the knowledge of the domain, the tasks of the domain, the interface they want to work with, and whether it is helpful to and usable by them. Also, it helps to determine whether it is acceptable to the managers and other stakeholders in terms of how it affects their tasks, their relationships with other workers and colleagues, the structure of the organisation, the management, communication with others, and so on.

Testing is conducted to detect mistakes and omissions in software, as well as to guarantee that the concept and framework of the programmes are suitable; it gives assurance that the software was produced and maintained appropriately. During this phase, tests appropriate for the code should be designed. Identifying all known and unknown paths is critical, then testing them.

Testing should be conducted of the upper and lower bounds (always test at zero), and several different types of testing should be employed. Testing methods include black box (input data with known and unknown results) and glass box (testing with full knowledge of the code, code walk-through, and standard audit). Acceptance testing is conducted to determine if the particular phase meets its established criteria, as determined by the project manager.

The final technical design of the system is determined. The design document should contain, for example, all the data to be used by the system, its format and its location; all the rules and knowledge to be included in the system; the user interface – including layout, communication format (e.g. whether it will be menu-driven), screen colours, and type of fonts; and interfaces to external languages and programmes.

After the system's design, the code is written using a programming language like SQL, Python, C, Go, Kotlin and JavaScript, or system shells. Acceptance testing should be performed during the design review to confirm that the design meets all of the criteria stated during the requirements stage. As is the case with the software system, the actual code should be developed with future needs in mind; that is, consideration of the needs of the maintainer to easily understand, test and modify the code. This can be done using a high-level language, standard programming conventions, standard compiler options, meaningful documentation and comments, and one common language for the whole organisation. During the life-cycle coding phase, avoiding bad habits that contribute to software problems and maintenance inhibition is imperative. The following are just some of the critical contributions to software problems:

- Deeply nested DO loops
- Unneeded code that has not been removed
- Excessive use of IF statements
- Excessive use of global variables
- Excessive use of GOTO statements
- Redundant calls and modules
- Lack of top-down design
- Lack of or no documentation (within the code)
- Overly complex programme structure and logic

The primary objective of implementation is the smooth operation of the software. Sometimes there will be a need to modify the system further. Final evaluation or impact assessments are not only to test and evaluate the technical functionality of the system but also to assess the impacts of the system on the users and the organisation. The final evaluation is done after several months of using the system. The last approach is introduced into the organisation and then evaluated after use. Management of change determines issues that need to be included in the implementation strategy for the system's deployment. It determines what needs to be done to ensure that the system diffuses into the organisation without resistance; and how best to prepare the organisation to use the system.

When a new system is introduced, it usually upsets the status quo. It might cause problems by disrupting set practices or culture or threatening some users or stakeholders. Systems are perceived to be formalised and standardised, and users see them as threatening, unneeded intrusions. Some domain experts see systems robbing them of their skills, job control and power. This calls for careful "change management" during the system's integration into the organisation. It is essential to understand how some stakeholders will react to the system; the introduction strategy should be planned so that they can see the importance and uses of the system to them and the organisation, realising that it will not take away their jobs, harm them or affect them deleteriously.

Also, training should be given to users where necessary. Maintenance activities are traditionally divided into three categories: adaptive, corrective and perfective or enhancement maintenance.

Adaptive maintenance includes all modifications made to meet the changing structure in which the software must operate. That is, data formats, hardware and operating systems change. It contributes to 20% of all software maintenance during the operation phase.

Actual errors necessitate corrective maintenance. The main reasons for corrective maintenance are logic and design errors. These modifications contribute 20% of all software maintenance during the operation phase.

Perfective or enhancement maintenance includes changes and alterations that occur because of changing or expanding user needs. This contributes to 60% of all the changes carried out during the operation phase. The modification could be in the form of corrective modification or maintenance of the system, for example, debugging incorrect code or enhancing the system's functionality. One can modify an existing system by correcting or changing existing functions, providing the system with additional functionality or a new interface, porting the system from one hardware to another, changing the system from a shell to a high-level language code implementation, or changing the operating system under which the system runs.

Software design practices should include criteria for maintainability, ensuring that simplicity outweighs completeness. Change management should include the limitation of the effects in the maintenance phase of a change made in the design phase and the determination of the ripple effects on other modules of making a change to a common module. This involves looking at global variables, modules which invoke or are invoked by a common module, as well as determining the effect on a module of making a change to a local variable. Design should be evaluated for excessive complexity.

Maintainability can be significantly improved if certain activities are carried out. In change management, implement the simplest revisions first, replace one module at a time, check suggested changes for every side effect, and conduct regression tests after every change. Create criteria for software modification and testing. For example, identify source comments that have been altered by assigning a reference number to the change request. Learn to read software programmes. Keep records of bugs and routine maintenance bugs. Variable declarations should be centralised in a programme. Use object classes to specify the correct types and values that objects of a kind can take and the set of operations that can be done on those types. Define a computer-processable database in a centralised, symbolically referenced manner. Because it is assumed that software will evolve and change, each developer in the planning phase should consider the next developer in the life-cycle.

Software maintenance should be a critical function in most organisations, so it is essential that the staff be highly skilled and motivated. The maintenance personnel should be experienced, versatile, analytical, adaptable and thorough. They must have a good application of skills and knowledge of various hardware, operating systems and languages. They must be able to find errors and understand how changing the code affects the entire programme. They must be able to work in challenging situations and under short timelines. Modifications and changes must be thoroughly debugged and tested.

To enhance the motivation of software maintenance staff, couple software objectives to organisational goals and maintenance rewards to corporate performance; integrate software maintenance personnel into operational teams; create a discretionary perfective maintenance budget, as well as perquisites of software ownership; rectify the negative image of software maintenance; integrate development and maintenance to obviate the situation wherein the two are treated as separate functions with different status.

References

Baxter, G. & Sommerville, I. (2011). Socio-technical systems: From design methods to systems engineering. *Interacting with Computers*, 23(1), 4–17. Retrieved May 28, 2022, from https://doi.org/10.1016/j.intcom.2010.07.003

Polson, D. (1990). *ORDIT: The development of tools to assist in organisational requirements definition for information technology systems*. Paper presented at the International Conference on Man, Technology and Organisations, Nivelles, Belgium.

Rosenbrock, H. H. (1987). Engineers and the work that people do. In *Information Technology: Social issues*, edited by R. Finnegan, G. Salaman, & K. Thompson. London: Southeast Asian Journal of Social Science, 17(2), 143–145. Retrieved May 28, 2022, from http://www.jstor.org/stable/24491150

Rosenbrock, H. H. (1990). *Machines with a purpose*. Oxford University Press, Oxford.

Thomas, M. (1991). *What constitutes a KBS methodology?* in Knowledge-based Systems Methodologies Workshop, Proceedings of the First SGES International Workshop, London, December 3 – 4 (British Computer Society Specialist Group on Expert Systems).

Woherem, E. E. (1988b). *Impact of expert systems in organisations: A case study*. Paper presented at the conference on Software Tools, July 14 - 18.

Woherem, E. E. (1991c). *Strategy for information technology manpower development at the organisational level*. Paper presented at the IFIP Workshop on Information Technology, Manpower Needs in Developing Countries, August 5 – 6, Ahmedabad, India.

Chapter 14

Strategy for Indigenisation of IT Systems in Africa

Over the past decade, IT investors have focused on Africa as a hub for significant IT investments and commercially oriented startups. Despite the continent's limitations, entrepreneurs and innovators are drawn to the potential of expanding technology to Africa's populations and economies. Africa's IT sector is representative of expatriate entrepreneurs with some of Africa's best brains. IT innovation is beginning to influence businesses of all sizes in Africa, including fashion, entertainment, agriculture, energy, education, healthcare, banking, and transportation. As a result, there is enormous hope that IT systems in Africa will have a more significant impact speedily than anywhere else globally. As Africa strives to move from the edge to the epicentre of the global economies, IT systems play an undeniably substantial role. Therefore, Africa needs indigenised IT systems that will overhaul regional trends in business, education, politics, infrastructure, investment and modernisation.

Indigenisation of IT systems is a process whereby African countries develop IT strategies that align with their national and economic development policies. They must implement an extensive set of design, process, and practice reforms that strengthen inclusion, tear down barriers and enable their sectors and populations to employ and leverage the benefits and possibilities provided by IT systems. Through the indigenisation of IT systems, African countries would be able to master, utilise, transfer, assimilate and properly manage and use IT systems to enhance the development of Africa. This will set the continent on the path of socioeconomic development, enhancement and technological exploitation.

So, this chapter examines the issues that are debilitating to the practical and successful acquisition, assimilation and indigenisation of IT in Africa and recommends some strategies for sustainable transfer, practice, and beneficial use of IT.

IT Acquisition and Adoption

First of all, IT acquisition can be defined as the process of purchasing IT systems. It could also mean importing IT from the countries of their invention to other countries. On the other hand, IT adoption means introducing, adopting, or localising IT to a new part of the world. So, IT acquisition and adoption refer to a country's complex processes and measures to buy and use IT inventions and systems.

IT acquisition and adoption are externally and internally driven. The IT knowledge can come from external countries but can be adopted internally through research and development. Different IT stakeholders are involved in both internal and external IT processes. Those stakeholders include the government, partners, investors, local and foreign IT experts, and end users. IT acquisition and adoption can be made through direct buying of items of technological equipment, licensing of the equipment, as well as technical support agreement.

Therefore, there is a symbiotic relationship between the world's developed economies and the Global South regarding IT acquisition and adoption. The Global North has attained a digitalised and knowledge-based economy characterised by a surfeit of availability of IT infrastructures and experts. Because developed economies have IT systems in large quantities, they have to look for IT markets in the world's developing economies. Developing economies, such as Africa, possess vast natural resources but a dearth of IT systems. Therefore, developing economies are the ones in need of IT acquisition and adoption.

IT acquisition and adoption are driven by individual, business, and governmental needs. Therefore, IT acquisition and adoption benefits individuals, governments and companies. With IT acquisition and adoption, there are chances of developing new products and gaining entrance into new markets. Technology acquisition and adoption are veritable tools to increase productivity in the developing countries of Africa. IT acquisition and adoption can increase a firm's efficiency, improve the quality of goods and services, increase consumer demand, and reduce production costs.

Acquiring and adopting already existing IT is a way of providing solutions to underdevelopment, unemployment and poverty. The developed solutions sought by the developing countries will be sourced abroad. For instance, Africa mainly benefits from IT infrastructures by importing readily available technologies from Europe and other developed countries to complement home-grown IT deficiencies. Note that IT acquisition and adoption are made by recognising the protection of intellectual property (IP) rights. Without the proper IP protection, the acquiring and adopting countries might be at legal loggerheads with foreign multinational IT firms or Original Equipment Manufacturers (OEMs). The latter will be forced to restrict research and development activities to benefit developing countries.

Challenges in the Transfer and Adoption of IT in African Countries

Now that African countries have embraced IT, which was birthed in Europe and other developed countries, they need to transfer and adopt IT to make it serve the continent better. However, as a continent in the Global South, IT transfer and adoption face critical challenges in Africa (Ejiaku, 2014). One of these challenges is the lack of requisite national IT policies. The IT policies of African governments are inefficient. This has hindered the creation and implementation of information technology. Even when governments develop IT policies, implementing them becomes a barrier. In addition, inadequate financing has often resulted in the abandonment or delay of IT projects.

Another challenge is the dearth of IT infrastructure. By IT infrastructures, we mean electricity, telecommunication, networks, computers, software, and hardware. Unfortunately, these IT infrastructures are not sufficient in Africa. The reason for this is insufficient investments in the IT sector. The funds which African governments allocate to IT infrastructures are insignificant. African states, therefore, depend on foreign aid for their IT infrastructure needs. Sadly, foreign donors' funds and other resources are ineffective in solving Africa's IT infrastructure challenges. This is because the funding from the foreign helpers is pre-packaged to solve some of Africa's IT challenges without enabling Africa to be IT independent in providing solutions for the local IT needs of the continent.

Away from infrastructure challenges, African countries lack enough skilled IT experts. They are the ones who can plan, design, build, install, and manage IT. The few available IT experts cannot succeed in adopting and transferring IT. The lack of commensurate IT professionals in Africa is government-related and educational. As for the governments in Africa, they have not given IT the proper attention. As for the educational system, our schools, colleges, polytechnics, and universities lack suitable environments for IT acquisition. For example, despite the Internet's widespread presence, African universities and polytechnic institutions have inadequate availability of the Internet and contemporary computing. Worse, the few available IT specialists in Africa still find it challenging to stay on the continent. Unfavourable conditions of service in Africa force many of them to migrate to Europe and other developed countries.

Moreover, political instabilities and insecurities in African countries affect IT adoption and transfer. It is of note that IT policies and infrastructures are valuable in adopting and transferring IT. Unfortunately, African countries like Somalia, Angola, Liberia, Sudan, Nigeria, and Guinea are home to political instability. IT policies cannot be sustainable in such volatile economies because frequent government changes affect them. Insecurities exacerbate political fluctuations in Africa. Many African countries currently house some of the world's most dangerous terrorist organisations. The implication is that no reasonable local or foreign IT investor will like to invest in areas of the continent battered by terrorism.

The Need for Indigenisation of IT in Africa

IT systems and infrastructures used in Africa come from Europe, the USA, China, and other developed economies. Therefore, there is a need for African countries to indigenise the IT systems and infrastructures used in Africa for the following reasons.

Africa has socioeconomic realities and cultures different from those of Europe and other developed countries. As there is a significant relationship between socioeconomic backgrounds, it is necessary for Africa to indigenise its IT systems and infrastructures. By indigenising IT in Africa, the continent will avoid the case of foreigners making Africa a dumping site for IT waste. With relevant IT indigenisation policies and actions, Africa will have IT systems and infrastructures to meet its socioeconomic needs (Hinrichsen, 2020). In other words, the indigenisation of IT in Africa will help the continent to meet the local needs of the people. It will also accelerate the digitisation of the economies of African countries.

Furthermore, Africa needs indigenous IT to address the IT infrastructural deficit on the continent. Africa still lags in IT infrastructures at this momentous stage of the Fourth Industrial Revolution. Software, hardware, Internet accessibility, broadband availability, etc, are not available in their proper proportions. Therefore, IT infrastructures need to be indigenised on the continent. Indigenisation of IT will encourage local IT experts and investors to develop IT infrastructures that can drive socioeconomic development on the continent. It will also give Africa a competitive advantage in the digitalised and globalised economies of the world and increase the number of local IT specialists and experts on the continent.

For IT indigenisation in Africa to be feasible, it must be led by indigenous IT experts. For instance, the participation of African IT experts will ensure the affordability of the price of manufactured IT equipment, components, and infrastructures, low-cost repair, and maintenance. Also, local IT specialists in Africa will ensure a periodic increase in IT infrastructures and systems by using indigenous upgraded technologies. Local IT experts are better positioned to anticipate and understand the technologies underlying such IT infrastructures. Therefore, they can change, modify, improve or re-design the IT infrastructures when needed. So, in effect, and as mentioned above, with the increase in Africa's IT experts, more jobs will be created for the teeming African populace, thereby reducing the perennial unemployment cases on the continent.

Strategic Framework for IT Indigenisation for African Countries

For Africa to close the IT divide between itself and the developed economies, some frameworks and measures need to be adopted. Therefore, I advocate the following frameworks.

Africa can indigenise IT by investing in IT research and development (IT R&D). This process should start by making IT-friendly national policies. After the policies, there should be government funds, grants, and technical support for local IT startups and investors across Africa. By investing in IT R&D, African countries can develop measures that will make them meet up with the demands of the 4IR. Meeting up with the needs of the Fourth Industrial Revolution is the surest way to achieve national development and global competitive advantage.

After national policies on IT, African countries should make IT infrastructures available. The first step in ensuring the availability of IT infrastructures in Africa is the provision of stable and reliant electricity. This is because electricity provides the enabling environment for IT infrastructures to thrive. After putting power infrastructures in place, the governments can further make provisions for national IT broadband infrastructures and Internet access, regulate reasonable prices for access to IT and increase their budgetary allocations for IT infrastructures.

Partnership with local IT technicians, operators, and investors is another viable framework for indigenising IT in Africa. The truth is that IT companies in Africa know the continent's IT needs and terrains. There are many mobile network operators in Africa, for instance, MTN, Safaricom, Airtel, Econet, 9mobile, Etisalat, Vodacom, and GLO. Partnership with these mobile network operators will help indigenous IT on the continent. This is because mobile operators can add plans and tariffs that cater to the IT needs of the local populace, even in the most remote areas, since mobile phone network coverage has penetrated most of the rural areas in Africa. In Tanzania, for instance, Jamii used Vodacom Tanzania's brand to increase health insurance penetration. The startup stated that Vodacom's recognised brand was essential in getting more individuals to pay for the micro-health insurance package.

Adapting IT to local needs is another strategic framework for IT indigenisation in Africa (Shekhar, 2018). Africans have needs and challenges that foreign-structured IT cannot meet. Therefore, indigenous African IT on the continent must re-structure IT infrastructures to fit local needs. For instance, at its inception in Nigeria, *FarmCrowdy* was using its online platform to encourage farmers and businesses to invest in farming projects. However, the company later discovered that its sponsors were interested in opportunities outside agriculture. Hence, *FarmCrowdy* had to look into possibilities that handled some other Sustainable Development Goals (SDGs). That was how the company added *CrowdyVest*. Currently, *CrowdyVest* addresses the other 17 SDGs, one after the other.

Indigenous IT education is paramount for Africa's move to IT indigenisation. The current realities orchestrated by the Fourth Industrial Revolution have changed educational needs and demands. Academic skills have undergone revolutionary changes from traditional to digitalised. People who get the most lucrative jobs today are those with soft skills in programming, business analysis, process modelling, quality assurance, app development, systems design, and maintenance. Unfortunately, Africa's education systems have not changed to capture these areas with more marketable skills. Therefore, African schools and universities must update their education and curricula to incorporate digitally-needed skills. The above soft skills should be indigenised in such a way as to reflect the special development needs of the continent. This will go a long way to prepare and equip Africa to reap the benefits of the Fourth Industrial Revolution.

References

Ejiaku, Samuel A. (2014). Technology Adoption: Issues and Challenges in Information Technology Adoption in Emerging Economies. *Journal of International Technology and Information Management*, 23(2), 59-68. Retrieved May 28, 2022, from http://scholarworks.lib.csusb.edu/jitim/vol23/iss2/5

Hinrichsen, S. (2020, September 25). *GSMA: Driving the adoption of tech solutions across Africa*. Mobile for Development. Retrieved May 28, 2022, from https://www.gsma.com/mobilefordevelopment/blog/driving-the-adoption-of-tech-solutions-across-africa/

Shekhar, S. (2018, June 2). *Need for a strategy for indigenisation of defence production – Part 1*. PGurus. Retrieved May 28, 2022, from https://www.pgurus.com/need-for-a-strategy-for-indigenization-of-defence-production-part-1/

Chapter 15

IT, the Fourth Industrial Revolution and the Future of Work

Information technology influences almost everything we do today. For instance, governments, businesses and individuals, in one way or the other, need its application and use. In this chapter, I am looking at IT within the context of the Fourth Industrial Revolution to see how it affects employment opportunities and societal decision-making.

The Fourth Industrial Revolution and Employment

The Fourth Industrial Revolution impacts employment and the work economy most profoundly. There is an increased demand for science, technology, engineering and mathematics (STEM) skills. With the exponential technologies riding the wave of the Fourth Industrial Revolution, individual learners can easily acquire STEM skills online. Online learning platforms have reduced learning barriers, and people can now develop educational skills and knowledge online, despite their location, once connected to the Internet. As citizens can learn new skills online, they can acquire emerging and relevant IT skills commensurate with the Fourth Industrial Revolution and reduce the risk of becoming redundant and unemployed because of declining skills (Schwab, 2016; Roy, 2019).

The Internet has opened numerous avenues for people to learn online. We have multiple massive open online courses (MOOCs) which have decentralised individual learning. Some of the notable MOOCs are Udemy, Coursera, and YouTube. With these online learning platforms, one can become a professional in one area within six months. The existence of these platforms means that Africans can use them to acquire the new tech skills they need today. In other words, individuals now have access to courses that teach them the relevant skills required in the 4IR era.

The nature of jobs and employment opportunities is changing due to the emerging skills of the Fourth Industrial Revolution (Roy, 2019). All around the globe, the gig economy is now the norm. By gigs, I mean flexible and temporary jobs that deviate from traditional work styles. In the gig economy, alternative work arrangements like freelancing and independent contracting are standard. Employers and companies are more comfortable hiring freelancers and independent contractors in a gig economy than full-time employees (Sandler, n.d.). With the digital technologies of the Fourth Industrial Revolution, work is changing as we once knew it.

The implication is that people with such relevant skills can be employed in any part of the world. For instance, in 2021, Jack Dorsey, the ex-CEO and co-founder of Twitter, appointed four Africans, Abubakar Nur Khalil, Obi Nwosu, and Ojoma Ochai (Nigerians) and Carla Kirk-Cohen (South African), to lead Bitcoin (BTrust) trust fund. This employment was made possible through the Fourth Industrial Revolution's gig economy.

The Fourth Industrial Revolution and the Future of Work – Labour Market Evolution

Labour market evolution looks at the changes and developments experienced in people's jobs. The changes could be seen in job locations, wages and salaries, relationships between employers and employees, and work arrangements. With the 4IR already pervading the job market, the future of jobs will experience dramatic changes (Choi, Dutz, & Usman, 2020). The changes are based on the insights we can get from the 4IR technologies.

Automation is the future of work. The insights from the current status of the 4IR technologies indicate that people's work will be dramatically automated. McKinsey Global Institute's Report concludes that half of the people's jobs could be automated by 2055(Manyika et al., 2017). Automation has some benefits. It enables organisations to boost productivity and profitability by reducing human error, working to improve performance and effectiveness, and sometimes performing tasks beyond human capabilities. With automation, tedious and routine jobs will be diminished, giving people time to engage in creative, collaborative, and complex problem-solving tasks that machine automation cannot handle.

However, will automation replace humans in the workplace? The answer is no, in at least the short to medium term. Automation will instead create a demand for more tech skills. In other words, the traditional skills employers value will be replaced with tech-savvy skills that align with the Fourth Industrial Revolution technologies. That is why, according to Schwab (2016), the founder and executive chairman of the World Economic Forum, talent, rather than capital, will be the critical component of production in the future. With this trend, he also predicts that people's jobs will be categorised into "low-skill/low-pay" and "high-skill/high-pay."

Besides, automation will create yet-to-exist jobs. The 4IR technologies will destroy some traditional jobs, but the good news is that they will create many new jobs that are yet to exist. Let us take a cue from what happened when computers were introduced. Computers replaced some jobs, for instance, those done by typists, but they created new jobs by developing, operating, and programming computers. So, as automation kills some jobs, it will make some in the future.

In all, work values will be disrupted in the future. In other words, what people used to value about work will no longer be in vogue. In the past, for instance, people could only vouch that you were working when you woke up in the morning, dressed and went to your place of work. However, now, that value has changed; some work from home. Imagine the future of remote work. The future of work is based on flexibility, as people's workspace, locations, arrangements, and relationships will be more of gigs and freelancing.

Accelerated Adoption of Technology in the Workplace

The emergence of the Fourth Industrial Revolution has led to the massive and accelerated adoption of 4IR technologies in the workplace. Therefore, this section discusses some areas that have witnessed accelerated adoption of the 4IR technologies in the workplace.

The first area here is the automation of production processes. Many firms and businesses have resorted to adopting exponential technologies in production. Such areas where technologies have been adopted include migration of assets to the cloud, increased spending on data security, and the use of the Internet, social media, and digital platforms. In addition, in response to the needs of the Fourth Industrial Revolution, companies have invested in new hardware, software, or technology solutions. These technologies are adopted to save time and cost, increase productivity and performance, and reduce drudgeries associated with manual labour.

Another area where there is accelerated adoption of the Fourth Industrial Revolution technologies in the workplace is human relationships. With the assistance of the Fourth Industrial Revolution technological innovations, the way humans interact and share information has dramatically moved online. For businesses and firms, interaction with customers is mainly done through online channels. We see AI technologies being used for customer service roles here and there. For example, many companies and firms today resort to using AI chatbots to deal with customer queries.

Moreover, the acceleration of Fourth Industrial Revolution technologies has altered work arrangements in the workplace. With the 4IR technologies, the age-old belief that you must show up for work at the workplace to prove you are working has changed. The incidence of COVID-19 brought the power of the Fourth Industrial Revolution technologies to make use of and understand that it is possible to work from anywhere, provided that your output is seen. In other words, we now have a gig economy where some companies, firms, and businesses contract job roles to freelancers (Sandler, n.d.). By using the Internet, gig workers can now work freely without the constraints and undue monitoring of the labour space.

Emerging and Declining Jobs

By emerging jobs, I mean jobs which are rapidly in demand due to the Fourth Industrial Revolution technologies. On the other hand, declining jobs are those being deskilled or displaced by the technologies of the Fourth Industrial Revolution. Therefore, the emerging-declining job dichotomy is a predictable outcome of the Fourth Industrial Revolution. See Table 2.

Table 2: The emerging-declining job dichotomy

Emerging Jobs	**Declining Jobs**
Data Analysts and Scientists	Data Entry Clerks
AI & Machine Learning Specialists	Administrative and Executive Secretaries
Big Data Specialists	Accounting, Bookkeeping and Payroll Clerks
Digital Marketing & Strategy Specialists	Accountants and Auditors
Process Automation Specialists	Assembly and Factory Workers
Business Development Professionals	Business Services and Administration Managers
Digital Transformation Specialists	Client Information & Customer Service Workers
Information Security Analysts	General and Operations Managers
Software and Applications Developers	Mechanics and Machinery Repairers
Internet of Things Specialists	Material-Recording and Stock-Keeping Clerks
Project Managers	Financial Analysts
Business Services & Administration Managers	Postal Service Clerks
Database & Network Professionals	Sales Rep., Wholesale and Manuf., Tech. and Sci. Products
Robotics Engineers	Relationship Managers
Strategic Advisors	Bank Tellers and Related Clerks
Management & Organization Analysts	Door-To-Door Sales, News and Street Vendors
Fintech Engineers	Electronics and Telecoms Installers and Repairers

Mechanics and Machinery Repairers	Human Resources Specialists
Organisational Development Specialists	Training and Development Specialists
Risk Management Specialists	Construction Labourers

Source: Schwab & Zahidi (2020) "The Future of Jobs Report 2020", *World Economic Forum.*

The issue of emerging versus declining jobs has implications for governments and people around the globe. On the part of the government, mainly in the developed economies, they handle redundancies and unemployment. As people lose their jobs because of their lack of 4IR-friendly skills, the burden of unemployment benefits is passed to the governments. To arrest such a situation, governments of the developed economies are further tasked with formulating policies that ensure their citizens upskill or acquire requisite skills commensurate with the Fourth Industrial Revolution.

Besides the governments, individuals all over the world feel the impact of emerging and declining job roles. The reality here is that employees who do not have the necessary Fourth Industrial Revolution skills are made redundant. For instance, robots can produce and finish factory products that take manual employees months to assemble and package within a day. If robots can accomplish such a task in a day and expenses are reduced, why employ humans who waste time and incur high costs? Nevertheless, employees will always retain their employability by possessing relevant skills matching the demands of the 4IR.

Emerging and Declining Skills

Emerging skills refer to skills increasingly in demand due to the emergence of the technologies of the Fourth Industrial Revolution; consequently, declining skills are gradually losing value courtesy of the Fourth Industrial Revolution. In other words, declining skills are no longer in demand in the global employment market. See Table 3.

Table 3: The emerging-declining skills dichotomy

Emerging Skills	**Declining Skills**
Technical skills and innovative thinking	Manual dexterity, resilience, and accuracy
Active learning and development strategies	Knowledge, verbal, perceptual, and spatial prowess
Innovation, creativity and entrepreneurial spirit	Quality control of financial and physical resources
Future tech architecture and programming	Digital maintenance and installation
Problem-solving abilities and modelling	Studying, writing, numeracy skills, and communication skills
Complex problem-solving	Management of technical staff
Governance and social power	System implementation and safety practices
Relationship management	Collaboration and organisation skills
Logical thinking, problem-solving and identity development	Conceptual, audiovisual, and dialogue attributes
System design assessment and analysis	Technology use, surveillance, and management

Source: Future of Jobs Report 2018, Web Economic Forum

For both employers and employees, emerging-declining job cases have some implications. Employers can give their current staff opportunities to reskill, upskill, or sack them to look for new employees. If employers choose to reskill or upskill the existing employees, their jobs will be spared. On the other hand, employers will exacerbate unemployment if they sack employees with declining skills.

Also, as old skills are giving way to new skills, employers of labour and business owners have to reposition their businesses to be 4IR-compliant. This is because the Fourth Industrial Revolution also impacts modes of running businesses. For instance, if a company used to have security guards before the emergence of CCTV cameras, the implication is that it has to upgrade by buying and installing CCTVs. This development is to the disadvantage of the low-skilled security guards.

Though business owners might embrace the technologies of the Fourth Industrial Revolution in their businesses, their attempts to adopt the underpinning technologies are hindered by emerging skills shortages. They might find it hard to attract suitable job candidates with the skills needed during the Fourth Industrial Revolution. This is because most job candidates have been trained in old curricula with traditional and declining job categories. Therefore, it is challenging for applicants with outdated skills to secure jobs in the Fourth Industrial Revolution economies.

As business owners keep embracing emerging skills, reskilling and upskilling are options left for the unemployed or those whose jobs are at risk because of a lack of the requisite skills. They have to go back and begin learning the skills relevant to the era of the Fourth Industrial Revolution economy. Their knowledge and acquisition of new skills are determinant factors that can land them jobs. Thanks to the 4IR-enabled learning tools and platforms, labour providers in need of emerging skills can quickly learn them online at their own pace.

Automation, Artificial Intelligence and Universal Basic Income

"Essentially, in the future, physical work will be a choice. This is why I think long term there will need to be a universal basic income."

– Elon Musk, the CEO of Tesla, in *Business Insider Africa, August 20, 2021*

Artificial Intelligence, automation, and Universal Basic Income have generated debates among investors, governments, researchers, and many others. AI is a field of computer science that builds intelligent machines to perform human tasks. By automation, we mean the application of programmed machines and technologies to perform duties previously done by humans. On the other hand, Universal Basic Income (UBI) represents a policy where all citizens of a country are entitled to a total amount of money every month, apart from the income they earn through other means. The source of UBI is the government and public institutions.

The advocates of UBI, like Elon Musk, say that AI and automation are rapidly taking over people's jobs. Moreover, if jobs are automated, many citizens will be jobless, leaving them with no sources of income.

The proponents of UBI advance some benefits of it. UBI will provide income for those whose jobs are replaced with AI and automation. In turn, UBI will alleviate poverty and save the government from social programmes that require more extraordinary bureaucratic processes. UBI will also empower caregivers and non-working parents. Citizens will have enough money under the UBI to cover living expenses, goods, and services. UBI will also reduce the rate of school dropouts and so on.

However, UBI is only accepted by some. The opponents of UBI have their counterarguments why governments should not implement it. According to them, UBI will discourage citizens from working and numb creativity, as many will lazily wait for the free money at the end of every month. Moreover, once citizens' enthusiasm to work is doused, the economy will be affected. In addition, UBI makes it hard for the government to help those who are most in need of help because both the rich and the poor are entitled to UBI, depriving the poor of their targeted support.

Though AI and automation are radically taking over people's jobs, other jobs will still emerge. We can already see some emerging jobs. Therefore, UBI is not an overall solution for jobs and skills that AI and automation are replacing. Instead, governments should provide enabling environments that can make it possible for declining jobs to be replaced by emerging jobs. That way, UBI will not be necessary.

References

Choi, J., Dutz, M. A., & Usman, Z. (2020). The Future of Work in Africa. World Bank Publications.

Manyika, J., Chui, M., Miremadi, M., Bughin, J., George, K., Willmott, P., & Dewhurst, M. (2017, January 12). *Harnessing automation for a future that works*. McKinsey & Company. https://www.mckinsey.com/featured-insights/digital-disruption/harnessing-automation-for-a-future-that-works

Roy, K. (2019, November 26). *The Fourth Industrial Revolution is redefining the economy as we know it*. World Economic Forum. Retrieved May 12, 2022, from https://www.weforum.org/agenda/2019/11/the-fourth-industrial-revolution-is-redefining-the-economy-as-we-know-it

Sandler, G. (n.d.). *The gig economy is going global*. American Express. Retrieved May 12, 2022, from https://www.americanexpress.com/en-us/business/trends-and-insights/articles/the-gig-economy-is-going-global/

Schwab, K. (2016, January 14). *The Fourth Industrial Revolution: What it means and how to respond*. World Economic Forum. Retrieved May 12, 2022, from https://www.weforum.org/agenda/2016/01/the-fourth-industrial-revolution-what-it-means-and-how-to-respond/

Schwab, K., & Zahidi, S. (2020, October 20). *Preface - The future of jobs report 2020*. World Economic Forum. https://www.weforum.org/reports/the-future-of-jobs-report-2020/in-full

Chapter 16
IT Strategy Planning

An IT strategy of an organisation is a blueprint for developing an information technology capability to support the organisation's business or corporate goals, create sustainability and maximise value return on investment. An IT strategy is implemented using an IT strategy plan. The step-by-step IT strategy process outlined in an IT strategy plan is known as the IT strategy planning process.

As technology becomes the bedrock of all aspects of business, organisations must create a holistic IT strategy that is integrated with and improves their business and organisational plans. The universal connectivity of the contemporary world means that each company must accept the reality that IT is no longer a support function. A robust IT strategy is a necessity rather than a luxury. A broad and all-inclusive IT strategy is crucial for a business to survive and prosper in the marketplace. It is imperative in today's era of exponential technologies. Organisations need to use good IT strategy planning to determine all the systems it would require over time to be competitive, including the new exponential technologies.

Key Elements of an IT Strategy

The key elements of an IT strategy include the following:

Mission and Vision: This is the first element of an IT strategy. A mission statement reveals what you should do and the IT systems that will effectively serve your customers. A vision articulates where your organisation desires to go regarding IT and how you can get there in a reasonable amount of time.

Roadmap: This is a detailed outline of the organisation's steps to institute a well-structured IT system from which it will derive maximum value and achieve competitive advantage.

Alignment with Business Goals: IT pervades every aspect of business, and it is foundational to the overall success of any organisation. Therefore, it is crucial to align IT and business goals. Knowing an organisation's objectives is essential for developing the IT strategy to achieve the goals.

Budget: Knowledge of the IT systems and processes you need to institute helps an organisation determine how much must be set aside as a budget for developing an IT strategy and implementing it.

Competitive Analysis: An IT strategy should include an analysis of competitors to determine the company's competitive edge over its rivals to zero in and strengthen its advantage.

Short and Long-Term Goals: An IT strategy incorporates short-term and long-term goals. Short and long-term goals should match the organisation's IT systems strategy. Short-term goals could be an assessment of the current IT infrastructure in your organisation or identifying the low-hanging fruits (or easy-to-design and implement IT systems), plus the IT processes in your organisation that are non-value adding or need to be improved upon. Examples of long-term goals could include full-blown IT systems that could be used to automate some achievable processes in about three to five years. Major exponential systems such as artificial intelligence, the Internet of Things, and blockchain technology could be among the long-term goals.

Continuous Assessment and Monitoring: By continually monitoring and evaluating your systems, you will always know the state of your IT systems and what improvements you might need to effect in the future.

IT Infrastructure: In developing an IT strategy, there must be an itemisation of all the software, hardware, applications and networks necessary to institute the system to run the business and achieve goals.

The Importance of IT Strategy

The IT function of an organisation is pivotal to its success because it underpins the operations of the organisation. Therefore an IT strategy designed to ensure the presence of the necessary IT infrastructure is essential in the following ways:

1. It removes or minimises operational bottlenecks, allowing the organisation to run more smoothly.
2. It gives leverage to achieve economies of scale and exact maximum value from IT investment.
3. It infuses agility into an organisation's operations and enables it to quickly respond to competition, changes in the marketplace and unexpected emergencies.
4. A robust IT strategy may provide a competitive advantage over competitors.
5. An IT strategy is essential for executing business strategies and managing risks.

IT Strategy vs Corporate or Business Strategy

Generally, an IT strategy is designed to capture the long-term plan of an organisation to create and continuously improve its IT capabilities. An IT strategy focuses on instituting an IT function that will support its operations to achieve maximum efficiency, profitability and sustainability. Usually, an IT strategy is aligned with a business strategy, with measurable indicators and a clear roadmap for its execution.

Since an IT strategy must be integrated with a business strategy, the initiative should be undertaken by a team comprising IT people, those developing the overall business strategy or those in charge of the strategic business unit. There should be input from all sides of the team to ensure that the IT strategy is fully aligned with the business strategy. With technology underpinning the operations of most businesses, integrating IT and business strategies is the surest way to achieve the organisation's goals.

The alignment of IT and business strategy confers various benefits on an organisation, including better allocation of resources, achievement of competitive advantage and sustainable business operations.

IT Strategy Planning/Developing an IT Strategy Plan

IT strategy planning is the process of developing plans that allow the organisation to leverage technology and technology-enabled services to achieve the objective of creating value for its stakeholders, investors and customers. It involves developing the "pathway" by which an organisation employs technology for desired business outcomes.

IT strategy planning starts with an assessment of the existing technology in the organisation, followed by a determination of the necessary capabilities to be developed to ensure sustainable operations and a definition of a roadmap and strategy for achieving the desired future.

The Benefits of IT Strategic Planning

A robust IT strategic planning offers the following benefit:

1. It makes it easier to delegate responsibilities. With a clearly articulated IT strategic plan, it becomes easier for the Chief Information Officer (CIO) to delegate duties to subordinates in the IT unit of the organisation. Once responsibilities are assigned, each task is executed based on the written-down strategic plan.

2. It provides a framework for making decisions on IT-Related matters. Since IT strategic planning usually includes the vision, objectives, requirements and processes, it provides a framework by which employees make decisions regarding the IT function of the organisation.

With a clearly-defined IT strategy framework, employees can make line decisions without necessarily involving management. This would make the organisation more agile and responsive to changes.

3. It helps an organisation to be agile, proactive and more responsive to change. Strategic planning equips an organisation with the capacity to quickly adapt to changes in the marketplace or industry.

4. It helps organisations prepare for contingencies. Strategic IT planning enables an organisation to respond quickly to emergencies without getting knocked offbeat. An IT strategic plan usually includes protocols triggered in an emergency to avoid catastrophic damage or loss to the organisation.

IT Strategy Planning Process

The IT strategy planning process is the step-by-step approach to creating an organisation's IT strategy. The IT strategy planning process has four phases: discovery phase, analysis and strategy formulation phase, implementation phase, and measurement and review phase.

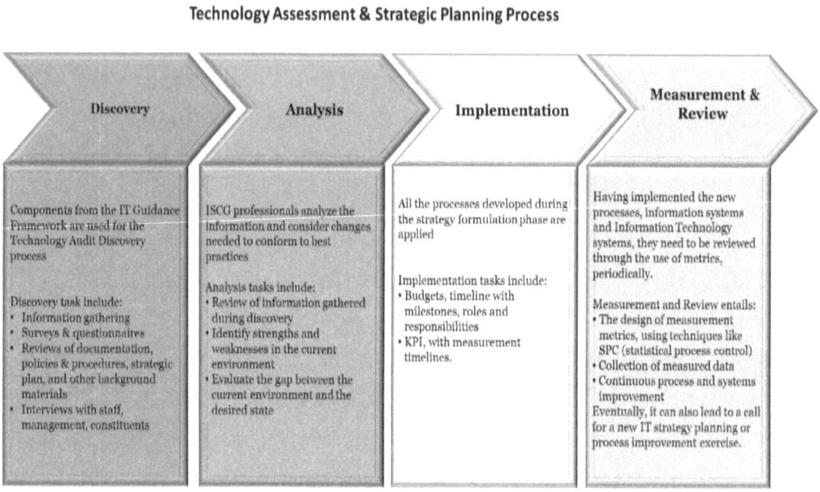

Figure 32: Technology Assessment & Strategic Planning Process

Discovery Phase: This is when the team leading the IT strategy planning process, including the Chief Information Officer (CIO), meets with management and heads of various business units to identify challenges and potential opportunities and obtain buy-in from stakeholders across the organisation. At this stage, the IT strategy team seeks to align IT strategy objectives with overall organisational goals and objectives.

Analysis and Strategy Formulation Phase: This is the second IT strategy planning process phase. The first part of this phase involves an analysis of the feedback gleaned from stakeholders. A SWOT (strength, weakness, opportunities and threats) analysis is conducted to determine the strengths, weaknesses, opportunities, and threats. Once strengths and weaknesses are known, opportunities to be exploited are identified, while measures are instituted to understand and guard against threats. Then a decision is made on which type of IT systems need to be built or incorporated to enable the organisation to operate optimally to achieve its corporate goals.

The second part of this phase involves defining the specific deliverables and outcomes to be achieved. KPIs, timelines with milestones, roles and responsibilities are outlined and adopted by everyone involved in the IT strategy planning process. The risks involved and steps for mitigating such risks are also identified. Then funds are allocated for the institution of the IT systems identified as offering the best promise to achieve its corporate goals.

Implementation Phase: Implementation is about executing the plan and managing everything to fruition. All the processes developed during the strategy formulation phase are applied, following the timeline and milestones captured in the plan. While implementing the IT strategy, a strict eye should be kept on any potential risk that could derail the project's timeline, budget or goals, and make plans to address such risks as quickly as possible in the event they arise.

Measurement and Review Phase: This is about evaluating the effectiveness of the IT strategy phase. This phase includes a feedback loop that allows the IT strategy planning team to determine where the plan needs to be revised or iterated for better results. The IT strategy planning process is an adaptable process that must be continually repeated to ensure that it provides the driving force that propels the organisation forward.

Traditional Approach vs Agile Approach to IT Strategy Planning

There are approaches to IT strategy planning. These are:

1. Traditional Approach: Under this mode of IT strategy planning, the IT strategy team – employees working in the IT department or external consultants – spends several weeks communicating with relevant stakeholders to get a good grasp of the organisation, business and current IT capabilities. After that, the team creates a strategic plan for the IT unit that extends over three to five years. This vision must include the capabilities incorporated into the IT systems to enable the organisation to achieve its goals. The capabilities are then broken into individual project lists or initiatives to be executed in a specific order. Finally, a road map for implementing the IT strategy is developed. The traditional approach to IT strategy planning delivers a new vision and a new set of objectives and targets every three to five years.

2. Agile Approach: The steps involved in the agile approach to IT strategy planning are the same as in the traditional method. However, in the case of the agile approach, after the initial development of a strategy plan, the team reconvenes every quarter or twice a year to review the progress made, discuss technology changes and industry trends, and adjust the road map accordingly. The agile approach keeps the road map fresh by ensuring that necessary corrections due to new developments and trends are incorporated into the strategic plan as soon as possible. Over the long haul, the agile approach to IT strategy planning offers more advantages than the traditional approach, primarily due to today's fast-changing world.

IT Strategy Framework

An IT strategy framework highlights the essential elements of an IT strategy plan that has been developed for the IT unit of an organisation. It explains how those elements should interact to enable the organisation to achieve the IT capabilities it needs to perform its corporate goals (CIO Wiki, 2022).

The IT strategy plan highlights the IT capabilities that an organisation should develop to further its business goals. At the same time, the IT framework is the conceptual structure that guides the development or integration of those capabilities. An IT framework incorporates the resources and relationships necessary for the optimal functioning of the IT strategy.

What is not included in an IT Strategy Framework

While an IT framework creates the structure that guides the development of IT capabilities, a few specifications are not included in the framework. These include the following:

i. It does not include a step-by-step sequence of the implementation of the IT strategy.
ii. It does not assign responsibilities to the individual members of the IT strategy planning team.
iii. It needs to specify the deliverables expected from the respective initiatives included in the strategy plan.
iv. It needs to provide a timeline to guide the implementation of the IT strategy.

Instead, an IT strategy framework usually would have a mechanism that:

i. Connects each IT strategy initiative to the overall organisational goal.

ii. Indicates a positive ROI before pursuing any IT strategy initiative.
iii. Explains the benefit of each initiative in quantitative terms, mainly where the benefits are not money-related.
iv. Determines that an initiative impacts the organisation's objectives before investing money.
v. Consistently monitors the impact of each IT strategy initiative on the overall goal of the organisation to determine if there is a need for course correction or otherwise.

5 Steps to Creating an Effective Information Technology Strategy Framework

The following five steps will assist in developing a practical IT strategy framework. It is essential to ensure that any IT strategy framework created is simple enough to be easily understood but effective enough to serve its purpose fully.

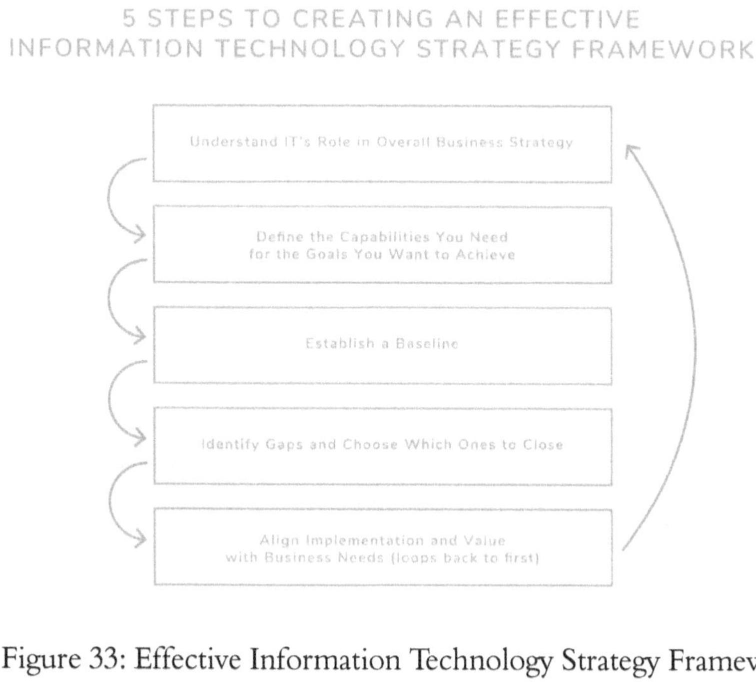

Figure 33: Effective Information Technology Strategy Framework

1. **Understand IT contributions to overall strategic planning.** The essence of an IT strategy is ensuring that the organisation has the IT infrastructure required to carry out its

operations and achieve its corporate goals optimally. Therefore, an IT strategy is never created for its own sake; it must be aligned with the overall business goal. To develop a practical IT strategy framework, a comprehensive understanding of the organisation's business and the IT capabilities needed to deliver value is required. A practical IT strategy framework should provide a guide for instituting IT capabilities for achieving organisational goals both in the present and future.

2. Define the capabilities needed for the goals to be achieved. This involves creating a map of the organisation's current IT capabilities to determine what exists and what needs to be incorporated to sufficiently equip the organisation to achieve its corporate goals.

3. Establish a baseline. The map of existing IT capabilities created should serve as a baseline for the capabilities required. Going forward, define the extent of the IT structure that must be in place to enable the organisation to achieve its goals.

Given the fast-changing nature of technology, it is imperative to consider that new software is constantly being developed and might require the IT strategy to be adjusted or course-corrected from time to time to make it more attuned to the organisation's goal.

4. Make a decision on which gaps to close based on the capabilities needed for now and the future. Proceed to conduct a gap analysis by comparing the existing IT capabilities with the infrastructure to be created. The study will expose the holes in the current system which must be catered for in the new IT strategy. Closing these gaps by incorporating or building the IT system needed usually requires skilled experts who could be people within the IT unit of the organisation or external consultants. The team will create the IT infrastructure to enable the organisation to move in the desired direction.

5. Merge implementation and value with organisation objectives. This is the final step in creating a practical IT strategy framework. It entails ensuring that each IT strategy initiative is tied to the value it will provide the organisation. Each deliverable must be able to help the organisation to achieve its present and future objectives. It should be clear from the framework that investing resources into implementing the IT strategy would not amount to throwing money down the drain.

References

CIO Wiki. (May 12, 2022). *IT strategy framework.* CIO Wiki. Retrieved June 22, 2022, from https://ciowiki.org/wiki/IT_Strategy_Framework

Chapter 17

IT and Corporate Boards

A board of directors is the governing body of an organisation that oversees management. Every public sector company is required to have a board of directors. In Nigeria, as in most other African countries, it is also necessary to have a board of directors.

Typically, the board of directors consists of the chief executive officer of the organisation, some other members of the management team and individuals who are not affiliated with the organisation. The composition of a board of directors is designed to ensure that there are people present on the board with expertise in the organisation's business, as well as those with different backgrounds and experiences who can each bring a unique perspective from what prevails in the particular industry. Shareholders appoint people to the board as either executive or non-executive directors. Board members meet regularly to deliberate and make decisions on issues about the organisation.

Using Technology to Achieve Good Corporate Governance

Technology and continuous innovation offer organisations with technologies to be more efficient, effective and productive. In like manner, technology has made a significant contribution to good corporate governance. New technological tools have transformed the board of directors' oversight function.

The areas where technology has had the most positive effects on corporate governance include the following areas:

1. Technology enhances transparency and accountability.

2. Technology enhances public participation in the affairs of the company participation.
3. Technology encourages effective public product and service delivery.
4. Technology provides tools to safeguard the organisation from cyber and other security breaches.

To ensure that technology is utilised correctly to assure safety from cyber security and other security breaches, an organisation needs proper management and technical expertise to manage technology and the right policies, procedures, and frameworks according to best practices.

Electronic Board Portals and Good Corporate Governance

In carrying out the activities of a board, organisations have realised and have started using numerous board technologies in the market, especially Electronic Board Portals (EBPs). Thus, EBPs are now as valuable for boards of directors of a company, just as electronic management tools are also vital for the operations of manufacturing companies.

Board portal software or EBP is an electronic system that supports board function by empowering management (board administrators) and board members to gather information, write reports, receive the board's feedback and directives, and generally manage most board activities. It is usually centralised, usable even from a smartphone in the hand of a director, and very secure, used to access board papers, schedule meetings, and execute goals. They can usually be accessed with the Internet from any place in the world.

The Key Feature of Board Meeting Software

Some of the critical features a board meeting software should have are as follows:

Document Management: A board portal usually has robust features in document management. The board app lets users add, edit, annotate, and track document updates.

Agenda Building: Agenda building is a crucial feature of board meeting software. The portal usually has a provision for users to create agendas for meetings. It lets you edit and collaborate on the meeting agenda to be more effective.

E-signatures: Board members may sometimes come together in a physical location for meetings. When meetings are held remotely, approving documents with the board members' signatures can be challenging. E-signing papers solve this problem and legitimise documents for further action.

Role-Based Permissions: Boards have various role players. Each of them has unique talents that they bring on board. Some of them play a role in different board committees. To facilitate the smooth flow of communication between notable stakeholders, it is essential to use board meeting software with role-based permission tools.

Voting/Polling: A board meeting portal should contain either voting or polling features to establish consensus among board members on hiring team members, management policies, and other board-related obligations. After the use of electronic voting, the system decides the outcome.

Secure Messaging: Board portals offer a secure channel for communication among board members. Such communication channels are more secure than channels such as email.

Usual Technological Solutions EBPs Bring

EBPs come with different technology solutions to ensure that boards work effectively and efficiently in carrying out their duties with accountability, transparency, and securely to ensure that good corporate governance is maintained. The following are some of the basic features of most EBPs:

1. Minutes Taking Function. The board secretary is responsible for minutes-taking at board meetings. Minutes-taking technology has dramatically simplified the secretary's duty of taking minutes at board meetings and getting approval on the minutes. With board portal software streamlining the minutes-taking task, the secretary is freed up to attend to the other responsibilities of the position, such as giving board members information or direction on matters about the board and answering their questions.

The use of the board portal significantly enhances the accuracy of the minutes. Since board meeting minutes could be summoned as proof, it is essential that the minutes recorded properly represent the decisions and actions of the board.

In addition, the security of board meeting minutes is paramount. Notebooks, loose-leaf papers and scratch paper could easily fall off a table or out of a file. With the electronic board portal, the security of board meeting minutes is guaranteed.

Board members need to readily get access to the last minutes as well as older minutes. The board meeting portal allows directors to pull up past board meeting minutes, read through and append their signatures appropriately. If necessary, the secretary may consider them, make the required corrections, and instantly transmit the updated minutes to all directors for approval with a single click.

2. Electronic Programme for Board Members to Complete Director and Officers (D&O) Questionnaires. On a yearly basis, board members are issued Directors and Officers questionnaires to provide information about their background and experience, securities ownership, conflict of interest, independence, compensation, and insider transactions.

The information supplied in the D&O questionnaire allows the organisation, through its lawyers, to provide necessary, accurate disclosures to the required regulatory authorities.

In the past, it was quite a task to have all the questionnaires completed and returned at the appropriate time with all the information required. The board secretary had to remind board members to complete the questionnaire repeatedly. And when the directors returned the questionnaires, some of the questions would not be answered correctly, and many fields would be left blank without the required information. Many board members considered the completion of the questionnaires to be mundane and unnecessarily laborious.

Board meeting portal technology has dramatically changed the dynamics. How organisations now create and distribute questionnaires have changed for the better. They no more contain redundant information, and no Director or Officer is sent questions that do not pertain to them. Moreover, Directors and Officers spend far less time completing the questionnaire, after which they submit with one click.

3. Electronic Board Assessment Tool. Generally, the performance of a board and its members must be assessed yearly to ensure they are continually competent and productive. Electronic Board Portals (EBPs) usually come with a board assessment functionality that aids in determining the board's general performance and the performance of each director. Directors usually must self-evaluate the board's capabilities and preparedness for the governance responsibilities they are saddled with. The areas of assessment include:

- Composition, diversity, skills, knowledge, experience and outlook
- Directors' independence
- Succession planning
- Efficient working of board committees
- Management supplies information to the board, where the questions arise: Is it done promptly and appropriately? Is the information supplied adequate, comprehensible, and reviewable?
- Quality of meeting and the nature of discussions. They should be open, forthright and constructive, and each board member's view should be taken into account without anyone dominating discussions.

4. Voting and Resolutions Tools. Electronic voting software for corporate board meetings is a technology tool used for voting and conducting polls in a secure and reliable manner. The software allows people to vote remotely from any location without gathering or using traditional paper ballots.

The electronic voting software offers various built-in capabilities to ensure no voting fraud. Anonymous voting ensures that each director knows the resolution they have voted for without the other directors knowing. Access restrictions ensure that only auditors have access to voting data. The process is structured to be efficient as votes are counted electronically, which helps to save time and resources.

5. Electronic Solution for Managing Any Entity-Related Information. Entity management is managing the vital information of a company's entities in an orderly system that the company can track and view at any point to ensure that all entities comply with the requirements of regulatory authorities. As an organisation grows in complexity and significance, so does its necessity for entity management.

Electronic solutions for entity management allow an organisation to create a central repository of all information about its entities. The entities, in the form of data and documents, are stored in the main repository and can be searched and viewed with only a few clicks. This helps management maintain a holistic view of all entities at all times. Besides, keeping the data in a centralised and digitalised format protects it from potential security breaches.

6. Technology Tools for Secure Corporate Board Communication. Over the past several years, new digital tools that allow board members to communicate among themselves and senior management have emerged to change the cumbersome and tacky approach to board communication.

Maintaining secure communication is highly needed for good corporate governance. Therefore, board members need digital tools that will afford them a level of security that will put their minds at rest when they exchange vital company information. An extra level of protection is crucial to safeguard from unauthorised access. Board meeting portals usually provide a platform for secure corporate board communication.

Overviews of the 10 Best Board Meeting Software/Portals

Here are the ten best board meeting portals and software (Reitsma, 2022):

1. BoardTrac: Boardtrac is a web-based EBP video conferencing document and management software. Documents can be added and edited and text highlighted. A director who is a member of two or more boards can use a single access log-in to access materials and functions on multiple boards without remembering several passwords for the different boards. Access to BoardTrac can be via smartphone, laptop, or desktop.

2. BoardPro: BoardPro has functions that enable users to arrange online board meetings. Meetings can be set based on issues or departments. BoardPro is very flexible and can quickly call up a database or word processor file for modification. It is also able to distribute the same to needed recipients efficiently. Also, board documents can be worked on by an individual user or many users concurrently. Meetings are digitally recorded and saved. Users can access files from wherever they are working and from any device and location.

3. BoardEye: BoardEye is another EBP and can easily be used to organise online board meetings. It has functionalities for managing meeting agendas, document management, emails, and data access management. It has a document management feature that allows for the addition and sharing of files, the addition of annotations to documents and highlighting text. It also enables directors to work concurrently on the same paper. The e-signature tool allows board members to sign and endorse documents quickly. BoardEye has an agenda-building feature that enables the user to add minutes which other board members can edit by adding or removing portions. Access rights and levels for each director can be created using the system.

4. Wrike: This offers role-based access to boards where files can be shared, meetings scheduled, and directors collaborate. There is an agenda management feature for setting agendas, tracking agenda items, setting agenda reminders, and assigning tasks. With the annotation function, users can highlight the critical points of a document. Also, it has a whiteboard feature that enables board members to collaborate and get input from one another in real-time. Wrike is versatile and can be interfaced with several other software, including Dropbox, Word, and Google Drive.

5. Boardable: It can be accessed on subscription as SaaS (Software as a Service). It is entirely web-based, ensuring that it can be accessed through any device, such as a smartphone, iPad, or PC. Meeting time can be proposed in the app, discussed, and a decision arrived at without the need for emails or phone calls. There are features for keeping track of scheduled meetings and agendas and voting in relevant polls. With the minute's feature, details of meetings can be documented and stored for future use. Tasks can be added and assigned right from the minute's records. Boardable has agenda-building templates to give a quick start to agenda building, which can be sent to directors in PDF format.

Boardable automatically interfaces with calendar apps, including Google Calendar, iCalendar, and Outlook.

6. BoardEffect: BoardEffect is used to arrange board meetings, amongst other functions. Board members can easily affirm their attendance by clicking on a proposed meeting date. It also has a tool enabling board members to vote on any issue and approve board agendas, minutes, goals, reports, and new members.

With BoardEffect, board meeting materials can be annotated on a mobile device. Updates can be provided continuously to keep board members abreast of the latest developments. Directors can remind themselves to remember specific tasks they need to carry out. Post-meeting follow-ups with various members can also be set up to re-engage them in the conversation.

7. CGS Boardworks: CGS Boardworks is another good EBP for organising and managing board meetings and other board processes and documents. It has a feature that users can add annotations to a document right from the console instantly. This enables collaboration in an online or offline configuration and accelerates any review process. In addition, the standardised voting and survey features of CGS Boardworks allow board members to vote electronically. They can also append their signatures electronically to approve documents or decisions.

Accessing the software on a tablet, mobile device, or PC is simple. It provides the same smooth user experience across devices, which makes it an excellent choice for big corporations with various departments. In addition, Apple Calendar is integrated with CGS Boardworks.

8. Parabol: Parabol is a free, open-source board meeting software for small firms. It allows members to provide anonymous feedback, a convenient feature for board evaluation of works, thus allowing members to continue improving as a team. Board members can be grouped according to their membership in the various committees to allow for deliberations before coming together for the general board meeting. Meeting summaries of the various committees can be created so everyone can stay in the loop.

Any director or board administrator can input the agenda before or after the meeting. After the meeting, the minutes are written together with the action points resulting from the board meeting for consideration as to the extent of their completion at the next board meeting. Parabol integrates with various other applications like Slack, GitHub, and Jira.

9. Boardvantage: Boardvantage is a board portal solution from Nasdaq, the renowned stock market index provider. One- or multi-day meetings may be set up with just a few clicks using its meeting management tools. Details of the meeting can be added, board members' attendance tracked, and files added, and once it is time, the session can be started remotely.

The document management component of Boardvantage assists in chronologically storing and managing all the documents and files of the board and all the company projects. Its search function can be used to find relevant documents quickly. Nevertheless, there are other features for collaboration, such as the possibility to post comments, exchange notes, build surveys and questionnaires, and tag particular people or groups.

10. Diligent: Diligent is another powerful EBP. It has unique corporate management and governance functionalities that enable directors and administrators to manage the production and distribution of board papers and other documents to all stakeholders in or outside the organisation and to ensure that good governance practices are maintained. Documents can be added, edited and removed via a centralised document storage space. Also, documents can be easily annotated and board meeting minutes added. Android and iOS devices, and PC, can access the product.

In addition to board meeting management, diligent boards support governance resources. For example, the board may use the data and analytics it provides about the business and the industry, as well as the curated news and director resources, to make decisions.

Blockchain as a Tool for Corporate Governance

Blockchain is a powerful technology for corporate governance assurance as it can promote trust. Let us look at the nature and benefits of blockchain technology and how it can be used to promote good corporate governance.

What Is Blockchain?

As explained in Chapter 4, blockchain is a powerful decentralised electronic ledger system that is shared by all users, as they are peer-to-peer, such that any changes made in any of its user nodes are reflected in all the other nodes at the same time, without the ability for anyone to delete or change what had been done and captured in the ledger, thus making its transactions and their histories to be immutable.

Blockchain and Corporate Governance

- It promotes trust and assures process integrity.
- It removes the risk of fraud by maintaining a tamperproof audit trail.
- It enables management and directors to concentrate on important things rather than worrying about process malfeasance.
- It makes company ownership more visible and trusted.

Opportunities and Applications of Blockchain in Corporate Governance

i. **Audit trail and Recordkeeping.** Blockchains ensure that correct records of the owners of the company are kept. This is very important as there often needs to be clarity about who owns what percentages of a company due to poor record keeping.

ii. **Entity management.** With blockchain technology, companies can ensure that their in-house records, per ownership of shares, will always be similar to what is filed in statutory registries.

iii. **Digital Identity.** In a blockchain, your identity is always trusted and immutable

iv. **For shareholders.** Blockchain makes the transfer of shares easy, quick and cheap.

v. **Intellectual Property.** Blockchain technology aids in eliminating data piracy and data corruption.

vi. **Property Ownership.** Blockchain uses an immutable time stamp format; therefore, it is an organisation's highly trusted record keeper.

vii. **Real-time accounting.** Financial statements are more highly trusted when produced using a blockchain-enabled system. Stakeholders that want to understand the accounting results of an organisation, therefore, do not have to rely on or have to ascertain that company managers are of high integrity before they trust statements produced by them, nor do they have to depend on auditors combing through the comments to assure that the figures are correct before they can trust them.

viii. **Smart Contracts execution.** These are programmes run on a blockchain system that automatically executes when all the agreed conditions are met. They do this automatically when all their requirements are reached. This means they can be used for contracts between people or organisations, be it to do with shares, properties, money, and projects, without needing a middleman.

ix. **Transparency.** Blockchains assure the overall transparency of an organisation, how it is being managed, and its results. This is critical for investors and owners who can always use the data on the blockchain to be knowledgeable about the organisation's actual performance.

x. **Voting.** Blockchain-based technology can ensure that corporate voting is correct to the last vote. In other words, every voter's voice counts, as each is included in the votes counted. When shareholders of a company realise that each of their votes counts, it will compel more company owners to go to the Annual General Meetings of the company to ensure that they participate in decisions to move the company forward and to ensure that important issues are voted for.

Blockchains use and store data with the help of sophisticated mathematical-based encryption techniques, making it highly impossible for anybody to manipulate any validated data that is part of one of their blocks. Moreover, blockchains leave an audit trail upon which any person can verify any transaction about what was done, who did it, and when it was done.

References

PwC. (2015). *Corporate governance in the boardroom: A practical perspective.* A PwC Point-of-View Paper. Retrieved June 12, 2022, from https://www.pwc.com/ee/et/publications/pub/corporate-governance-in-the-boardroom.pdf

Reitsma, T. (2022, January 2). *10 best board meeting software for board management online.* People Managing People. Retrieved June 12, 2022, from https://peoplemanagingpeople.com/tools/best-board-meeting-software/

Chapter 18

National IT Development and Management

The power and influence of technology in our modern world are beyond imagination. Every nation aspires to be ahead in terms of technological advancement. Developed countries are so-called because of the high technological know-how they possess. World powers like the USA, China, Russia, France, Germany, and others, have achieved significant technological development. Undoubtedly, a nation's IT development level determines its position and political dominance in the world.

However, IT development in nations across Africa is yet to reach the stage where it will bring the desired results. Consequently, there is a need for the countries on the continent to plan their IT development and management strategies thoughtfully. This section will generalise Africa, using Nigeria as a case study.

The Current State of IT in Nigeria

Nigeria's IT sector has been an integral part of the growth of the nation's economy. According to the Nigerian Communications Commission Media Team (2021), IT contributed 12.45% to Nigeria's GDP in the last quarter of 2020. Since the outbreak of the COVID-19 pandemic in Nigeria, no other sector of the economy has been this productive. According to the National Bureau of Statistics of Nigeria, by the second quarter of 2022, the percentage contribution of IT to the GDP of the country was 18.44%. IT, therefore, is a net contributor to the GDP of most African countries.

One of the reasons for the IT growth in Nigeria is the availability of mobile phones in the country. The growth of mobile phone use in the country has remained impressive. Most Nigerians have mobile phones for calls, SMS, and browsing. According to Ceci (2021), 48.12% of Nigerians in 2021 had access to the Internet via mobile devices, and by 2026, it is projected to grow to 659.7%. The use of mobile phones in Nigeria cuts across all sectors of the economy: agriculture, health, transportation, governance, commerce, and entertainment. Mobile phone usage contributes significantly to the country's GDP in all these sectors. For instance, buyers and sellers can easily transact business online in agriculture using local platforms like picky.ng, Farmsquare, Agro Market, and even international platforms like AgriBros Market, Tradeford, KisanMandi, EC21, and others. Also, the availability of mobile phones has reduced the cost of browsing the Internet in cybercafés.

Internet connection has been on the increase in the country. Many Nigerians have gotten jobs via the Internet. The statistics from the Nigerian Communications Commission Media Team (2021) revealed that 147 million Nigerians are on the Internet, most of whom are into one online business. It also indicates that as of May 2021, Nigeria recorded about 76 million subscriptions on broadband (penetration of 40%) and 187 million lines in the voice segment. These figures represent 97.9% teledensity.

Another IT driver of the GDP is fintech services. In Nigeria, the fintech industry is booming. The increased use of mobile phones fuels the achievements. Nigerians are now enjoying the benefits of the Federal Government's cashless economic initiatives. With their phones, mobile banking has gained significant growth. Nigerians can use their phones for USSD or mobile app banking. Numerous fintech services help Nigerians pay online, take loans, and save. Some fintech services are offered by Carbon, Paga, Flutterwave, Paystack, OPay, and many more. And in the last lap of 2021, the Federal Government launched its e-payment app called e-Naira (Central Bank of Nigeria, 2021).

However, the current IT state in Nigeria is not all positive: there are many challenges bedevilling the IT sector in Nigeria. To tackle these problems effectively, we must first identify the obstacles.

Challenges Facing the Development of IT in Nigeria

The development of the IT sector in Nigeria has many challenges that confront it. Some of these challenges are listed and discussed below.

Over-dependence and Focus on the Oil and Gas Sector. The government puts a lot of attention and reliance on crude oil and its products, which has affected other economic sectors, including the IT sector. While it is inarguable that significant national revenue has come from the sale of crude oil and its products, it should be noted that the reason for this is that the government has neglected other sectors of the economy which would yield similar or even higher income if given the same attention as the oil and gas sector. Due to its oil dependence, Nigeria needs to provide IT with the critical attention that will bring about more remarkable IT development and contribute to its revenue.

Inadequate IT Facilities. IT thrives in an environment where there are adequate facilities for it. However, in Nigeria, the development of IT is hampered by inadequate facilities. The software, hardware, optical fibre, bandwidth, and other infrastructural accessories that can bring about the desired effects still need to be improved in the country. For instance, the bandwidth in most parts of Nigeria is still based on 3G or 4G networks. This leads to poor Internet connectivity.

Cybercrime. Nigeria is one of the countries notorious for cybercrime. Many Internet fraudsters are locally referred to as "Yahoo Boys." They dupe Nigerians and non-Nigerians using voice phishing, SMS, phone calls, emails, and social networking sites. The activities of cyber criminals adversely affect IT development in Nigeria. For instance, local and international IT investors and companies are wary of Nigeria because of cybercrime and avoid situations where much money is spent to ensure the security of a company's data.

Limited IT-Skilled Personnel. There are few highly IT-skilled personnel in Nigeria to operate complex IT facilities. This is despite Nigeria having some of the most prominent IT personnel in Africa, many of whom are in high demand in Nigeria and across the world. The number of existing IT-skilled personnel in the country is not enough for the country's requirements. Yet, many abscond from the country to work overseas, whilst many others work remotely from Nigeria for foreign companies. This sad situation makes the country's position precarious concerning the availability of IT-skilled personnel. This means that the country still needs to seek the assistance of skilled foreign personnel or train many new ones to work in its IT ecosystem. Worse still, many Nigerians interested in learning and improving their skills in IT are not acquainted with the standard facilities required for such learning. This is why most people who have completed IT training in Nigerian tertiary institutions are still far behind citizens of developed countries who have attained similar IT degrees. As long as Nigeria's IT facilities are substandard and inadequate, there cannot be a significant development of IT in the country.

High Cost of IT Facilities and Personnel. Getting IT facilities, skilled personnel, and IT training can be prohibitive. Most Nigerians, for instance, the unemployed and dependents, find it hard to afford IT facilities and personnel. There is also a high cost of data. With proper funding from the government, private IT firms and individuals may be able to afford standard facilities and employ highly trained personnel who can do the job effectively. As a result, private firms and individuals tend to settle for IT systems and facilities that they can afford, which in most cases are usually substandard.

Inadequate Power Supply. Poor power supply in most parts of the country affects IT facilities and resources. This is because most IT gadgets use electric power to function. People using IT equipment are often left stranded after the systems' batteries run down. Some users have taken a step further in searching for alternative methods to power their gadgets. This has brought about using solar systems, generators, and inverters. But not everyone can afford these alternatives because their purchase and maintenance can be very costly. The best solution remains an adequate supply of power throughout the country.

Everything above regarding Nigeria as a case study equally applies to other African countries, from Ghana to Kenya, Egypt and South Africa. The situation is even worse in many other African countries.

A Case for Developing National IT Policies and Plans. National IT policy can be defined as an official statement that specifies the objectives, goals, principles, and strategies that guide and regulate the development, operation, and application of IT in a country. On the other hand, a national IT plan refers to the intended action committed to achieving a nation's objectives, goals, and strategies for IT development.

Nigeria has always come up with national IT policies. There was one in 2001 and another in 2012. Currently, the country uses the National Digital Economy Policy and strategy plan developed in 2020 to guide its IT planning up to 2030. This latest policy aims to build a Digital Nigeria, which means an economy driven by digital technology, towards evolving a knowledge economy, and not necessarily an economy fuelled mainly by a natural resource like crude oil. It also means that it can assist the country in climbing to a befitting level of digital maturity.

The importance of IT policies and plans cannot be overlooked. The above policies are applauded. However, they can benefit from being developed further by assessing their current achievements and setbacks and then updating them to accommodate new trends and developments in IT.

So, why are IT policies and plans necessary in Nigeria? In the first place, IT policies and procedures are developed in Nigeria because, when adequately executed, they create an enabling environment for IT networks, operations and services, which boosts national development in all sectors of the economy. Besides, I always make a case for appropriate IT policies and plans for Nigeria because they will make Nigeria relevant even on the international IT scene. As long as Nigeria has workable IT policies and plans that are appropriately implemented, it will be easier to make the country become a net IT contributor to other countries of the world. Nigeria's IT policies and plans will also help it to be able to receive and benefit from global IT solutions.

IT empowerment is another benefit of having IT policies and plans in Nigeria. IT provides citizens with relevant skills at both national and international levels. So, if Nigerians acquire good IT skills, courtesy of good IT policies and plans, it will give them an advantage in getting jobs both within and outside the country. IT is a source of employment for many people in Nigeria already. It has brought about opportunities for job creation, wealth generation and distribution. It has even helped create numerous opportunities for other sectors of the economy. If Nigeria is to reduce unemployment, then the development of IT policies and plans is not negotiable.

The overall argument is that if Nigeria is to achieve rapid and radical socioeconomic development in the IT age, it must continue formulating and implementing sound and workable IT-based policies and plans. This is because the application of IT is holistic; IT is used in all sectors of the economy: agriculture, health, commerce, finance, governance, and others. With economic growth, the overall living standards of Nigerians will improve, and poverty will significantly reduce. Life expectancy will improve, and Nigeria's brain drain will reduce.

Guidelines for the Development of Integrated National IT Frameworks, Policies, Plans and Structures

In the previous section, we have seen the importance of national IT policies and plans. We must provide guidelines for developing integrated national IT frameworks, policies, plans and structures. National policies on IT development in Nigeria can be based on the following outputs: IT frameworks, IT policies, IT plans, and IT structures.

The four elements work hand-in-hand to achieve the goal of national IT development. IT framework is responsible for setting up the agenda that guides the development of other parts of IT development. IT policy depends on the IT framework and aims to provide the details of crucial policy commitments and considerations pursued by the government. The IT plan provides details of the programmes and initiatives for implementing the government's policy commitments regarding IT. And lastly, IT structures refer to all the institutional set-ups, such as the relevant national coordinating structures, bodies, or institutional arrangements, which support the formulation, development, and implementation of IT policies and plans. These elements are examined below to see the relevant guidelines for Nigeria to follow in its IT development journey.

IT Framework Guidelines

Developing a nation's IT policy, plans, and structures should start with workable IT frameworks. The national IT frameworks guide the development of the other elements. Thus, it provides the guide for an analytical framework for IT policies and plans. For this to work, the analytical framework should review, analyse, and appraise the Federal Government's socioeconomic development frameworks, such as IT policies, programmes, infrastructures, and the degree of IT deployment, utilisation, and development in the country. This analytical framework will help the government to define and specify the general problems and policy areas that need attention.

Nigeria's IT framework should review the national socioeconomic development policies, programmes, and long-term development plans to achieve national IT sector success. For instance, the IT framework will showcase how each segment of the IT plan can align with pertinent development visions of the country.

Besides, the IT framework identifies and crystallises specific IT developmental challenges. It does this by establishing the relevant socioeconomic development, vision, missions and strategies which can be applied to address IT developmental challenges.

Summarily, the national IT framework explores the potential available for Nigeria to transform its economy into an information and knowledge-based society and economy. This can be done by identifying specific institutional arrangements and structures for facilitating national IT policies and plans.

IT Policy Guidelines

Guidelines on Nigeria's IT policies are needed to:

- Address IT development challenges
- Create an IT-friendly environment
- Develop IT Human Resources
- Deploy and harness underlying IT potentials and opportunities.

Challenges plague Nigeria's IT sector. Therefore, dynamic IT policies must be formulated to address the challenges. These policies ought to ensure an enabling IT environment to facilitate IT deployment, utilisation, and leveraging for national development and economic growth (Dzidonu, 2002). The enabling IT environment should include socioeconomic and political dimensions. Areas of consideration here should consist of provisions of IT infrastructures, special budgetary and tax instruments, incentives, encouragement of private-sector investment in IT, legal and regulatory frameworks (such as IT copyrights, data protection and security, freedom of access to information, and cyber laws/crimes).

For First World-level IT development to be possible in Nigeria, there should be packages for IT human resources. The development of national IT human resources can start from school. For instance, Nigeria can implement STEM education that will encourage the acquisition of IT capacities that meet the changing demands of the digital world. For citizens who have left the school system (graduates, adults, etc), the Federal Government can develop IT resource programmes to encourage training and re-training of IT experts and personnel.

Finally, there should be IT policy guidelines for deploying IT opportunities and potentials to meet the needs of the national economy. The deployment should be comprehensive; it should cover all sectors of the economy, such as the national health delivery system, commerce and trade, education, agriculture, etc.

IT Plan Guidelines

Guidelines on Nigeria's IT plans are essential for national IT development. So, they are expected to perform some functions.

First, national IT plans should be geared toward realising Nigeria's socioeconomic development visions (e.g. Vision 2050). The IT plans should follow the missions and the strategies formulated to achieve the national socioeconomic visions.

The national IT plans should be time-bound. For instance, a long-term IT development framework could be divided into short-term plan phases, each having a fixed timeframe and all the steps following sequentially. Dividing Nigeria's IT plans into phases will make them practical, realistic, and implementable.

In addition, Nigeria's IT plans should be flexible enough. Flexibility will ensure modification, revision, and adaptation as the need arises during implementation (International Trade Administration, 2021). One of the ways to provide flexibility in national IT plans is to avoid going into specific implementation details relating to IT programmes and initiatives. Instead, for each IT programme and initiative, details should be developed and worked out during the actual implementation of the IT plans. This will make it possible for specific circumstances, constraints, opportunities, and developments to be considered. Nigeria's IT plans should address the cost, budgetary, and resource requirements and allocation and mobilisation implications of IT programmes and initiatives.

IT Structure Guidelines

To achieve desired IT development, Nigeria should:

- Set up national IT bodies and institutions to support IT policies and plans.
- Ensure top-level political leadership structures that enhance IT policies and plans.
- Institute legal frameworks for the national IT structures.

As national IT development cannot occur without relevant institutions, the Federal Government needs to set up relevant IT bodies. The institutional bodies should include individual experts, teams, boards, and task forces. These bodies should be monitored and controlled by national coordinating structures, for instance, the National Information Technology Development Agency (NITDA) (National ICT Policy, 2012).

However, the relevant IT structures cannot function without an enabling political climate. Therefore, favourable political leadership is needed. In other words, top-level transparent political leadership is required for national IT policies and plans to work. After formulating IT policies and plans, the bodies in charge of implementation must be allowed to do their jobs without unnecessary political interference. This does not mean there should not be proper regulations of the national IT structures. The Presidency, in conjunction with the Federal Ministry of Communications and Digital Economy, through the agency of NITDA, should provide the necessary regulations and legal frameworks to oversee the responsibilities and activities of IT structures in Nigeria. That way, accountability and success in IT national development can be achieved (FME. 2019, National ICT Policy, 2012).

The Coverage and Components of the Integrated National IT Frameworks, Policies, Plans & Structures

Integrated national IT frameworks, policies, plans, and structures involve many areas of the economy. Therefore, for there to be national IT development in the country, the following components should be applied:

- IT Human Resource Development (IT HRD)
- STEM Education

- IT Research and Development (IT R&D)
- Investment Climate for Mobilisation of Financial and Technological Resources
- IT Infrastructures
- Development of Standards, Best Practices, and Regulations
- IT-compliant National Civil Service

IT Human Resource Development (IT HRD)

It is one thing for the government to make IT frameworks, policies, plans, and structures; it is another to have the availability of IT experts and personnel that will help to execute them. That is why the success of national IT development needs IT human resources. By IT human resources, we mean human skills, initiatives, intelligence, knowledge, and potential required for the IT sector. They include IT software engineers, robotics engineers, AI specialists, blockchain specialists, telecom engineers, web developers, support technicians, and other technical specialists, especially in exponential technologies.

In other words, the challenge confronting the development of IT in Nigeria is the lack of IT resources. Therefore, there is a need for IT HRD. By this, I mean training and re-training of IT human resources. To succeed in pursuing national IT development, the government, through the relevant agencies, should provide IT HRD guidelines and ensure that they are executed. With adequate IT HRD, it would be easier to implement and realise national IT frameworks, policies, structures, and plans.

STEM Education

Education is one of the best institutions for Nigeria to realise its IT frameworks, policies, plans, and structures. Therefore, the Federal Government should implement IT-friendly educational strategies through the Federal Ministry of Education.

Investing in science, technology, engineering, and mathematics (STEM) education is one approach to this. Investing in STEM has other demands on our education curricula. In other words, our education curricula should be overhauled from basic to tertiary education levels to accommodate IT courses and skills relevant to the IT age (FME, 2019). Also, the government can strengthen IT and STEM education by setting up a special trust fund, establishing IT research centres in educational institutions, and training and retraining IT personnel.

IT Research and Development (IT R&D)

IT R&D is responsible for developing innovative activities to improve IT skills, infrastructures, and human resources or develop new ones. In other words, IT R&D conducts and supports cutting-edge research into new ITs. It also provides relevant data to support the development of national IT frameworks, policies, plans, and structures. Therefore, while formulating national IT frameworks, policies, plans, and structures, national IT R&D is sacrosanct.

It is noteworthy, however, to acknowledge that IT R&D exists in Nigeria. For instance, the Nigerian Communications Commission (NCC) has an R&D department. It provides grants and support for IT research across tertiary institutions in Nigeria. Other agencies like NITDA and Ministries like the Ministry of Communications and Digital Economy and the Ministry of Science and Technology should also give yearly research grants and awards to researchers in tertiary institutions and other research bodies. Therefore, adequate provisions should be made to help it achieve its objectives.

Making provisions for IT R&D involves many things. Firstly, the government should set up a national IT R&D centre. Secondly, it should empower IT R&D by making adequate budgetary provisions for its human resources and infrastructure.

Investment Climate for Mobilisation of Financial and Technological Resources

To succeed in national IT frameworks, policies, plans, and structures, the Federal Government should create an investment climate that can accommodate financial and technological resources. Many things are involved in this. Firstly, special tax packages, instruments, and incentives should be provided to promote national IT development. Providing these would require both public and private financing. In addition, they can be shared between the public and private sectors.

In addition, the Federal Government should encourage indigenous content development and adoption. Incentives and special considerations should be given to private sector participation in national IT development.

IT Infrastructures

With IT infrastructures, national IT frameworks, policies, plans, and structures can only be possible. We mean all the components required for national IT development by IT infrastructure. IT infrastructures include hardware, software, networking components, operating systems (OSs), servers, data centres, PCs, Internet connectivity, and power supply.

However, one of the major problems limiting the development of IT in Nigeria is the lack of adequate IT infrastructure. For instance, we have slow Internet connectivity, and people in rural areas have little or no access to Internet connectivity; power challenges do not allow those with slow Internet access to make progress. So, the government should provide the necessary IT infrastructures for considerable national IT development in Nigeria.

Development of Standards, Best Practices, and Regulations

IT best practices are guidelines, procedures, or methods applied globally in the IT industry that result in positive outcomes. While I am not advocating adopting European IT best practices in Nigeria, hook, line, and sinker, it would be wise to study the First World economies and learn from their IT best practices.

Learning IT best practices are not enough, however. There should be specific provisions, applications, and adoption of the practices. The best practices should cover IT climate, human resources, and infrastructures. If such best practices are implemented in Nigeria, whether in urban or rural areas, Nigeria will achieve universal access to IT. However, best practices need to be regulated. For example, the government should create and fully implement cyber laws for national IT development frameworks, strategies, plans, and structures, to be successful. Said laws will protect individual, business, corporate, and government data. The cyber laws will also prevent cybercrimes across the country. The Standard Organisation of Nigeria can superintend over the implementation and regulation of the adoption of standards and best practices in the Nigeria IT ecosystem.

IT-compliant National Civil Service

The Nigerian Civil Service comprises professionals and experts who implement the policies and programmes of the executive arm. Implementing IT policies are part of the functions of the Nigerian Civil Service. But do we expect implementing IT frameworks, policies, plans, and structures to be effective when most Nigerian civil servants are not IT-compliant? Therefore, governments at the federal and state levels should ensure that Nigerian civil servants are given the necessary IT environment, training, and facilities. These will encourage the use of IT facilities by civil servants. It would be required for the government to provide IT facilities and services for civil servants and also use IT instructors and experts to educate the civil servants on using IT tools and facilities.

Achieving National IT Development in Nigeria through the Knowledge Economy

The knowledge economy (often referred to as the Digital Economy) is a system of production and consumption dependent on intellectual capital. In the knowledge economy, goods and services are mainly based on knowledge-based tasks and processes. In other words, it consists of people's scientific discoveries, applied research, and intellectual property. In the knowledge economy, intellectual capabilities are valued more than natural resources or physical labour. It means de-emphasising natural resources as the mainstay of the economy and emphasising even more of those endeavours that use intellect and knowledgeable practices.

The knowledge economy is more fully obtainable in the world's developed economies. Countries with a more comprehensive knowledge economy are also those that have climbed higher in digital maturity, that is, to strategic or exponential levels of digital maturity (See Chapter 9). The operations of the knowledge economy are based on scientific discoveries and R&D. The knowledge economy also provides technical support and consulting. Other characteristics of the knowledge economy include innovation, incentives for entrepreneurship and knowledge, availability of skilled labour, etc.

However, some implications accompany the knowledge economy. The knowledge economy produces redundant employees who do not have emerging IT-related skills. In other words, the knowledge economy creates unemployment for unskilled employees. In addition, the knowledge economy incurs expenses for companies, as they have to invest in on-the-job training for their employees who lack the relevant skills. Nevertheless, countries must continually determine how to cater for their less digitally-skilled citizens so they are not permanently unemployed or unemployable.

Notwithstanding the above negative implications, the knowledge economy has many benefits. It generates numerous job opportunities across all sectors, leading to wealth generation and redistribution, rapid economic development, prosperity, and readiness for global competitiveness.

Studying the benefits of the knowledge economy in developed economies would be necessary for Nigeria to embrace it. Currently, Nigeria is still at the level of the natural resources-based economy, in which the bulk of the national budget comes from crude oil. It can, however, circumvent the classical agricultural to industrial to information development routes by leapfrogging development via the digital or knowledge economy route. In other words, embracing the opportunities offered by the information and knowledge economy will help the country leapfrog to the status of a developed nation.

References

Nigerian Communications Commission Media Team. (2021, February 22). *Press statement: Telecoms sector, others take Nigeria out of recession as sector boosts GDP by 12.45%*. Nigerian Communications Commission. https://ncc.gov.ng/media-centre/news-headlines/960-press-statement-telecoms-sector-others-take-nigeria-out-of-recession-as-sector-boosts-gdp-by-12-45

Ceci, L. (2021, August 24). *Nigeria mobile internet user penetration 2026*. Statista. Retrieved June 22, 2022, from https://www.statista.com/statistics/972900/internet-user-reach-nigeria/

Central Bank of Nigeria. (2021, October 25). *eNaira*. Central Bank of Nigeria. Retrieved June 22, 2022, from https://www.cbn.gov.ng/currency/enaira.asp

National ICT Policy. (2012). National Information and Communication

Technology (ICT) Policy. NITDA. Retrieved May 12, 2022, from https://nitda.gov.ng/wp-content/uploads/2020/11/National-ICT-Policy11.pdf

Dzidonu, C. (2002). A blueprint for developing National ICT policy in Africa. African Technology Policy Studies Network. Retrieved May 12, 2022, from https://atpsnet.org/meki/wp-content/uploads/special_paper_series_5.pdf

International Trade Administration. (2021, October 13). *Nigeria - Information and Communications Technology*. International Trade Administration. Retrieved May 12, 2022, from https://www.trade.gov/country-commercial-guides/nigeria-information-and-communications-technology

FME. (2019). National Policy on Information and Communication Technologies (ICT) in Education. Federal Ministry of Education. Retrieved May 12, 2022, from https://education.gov.ng/wp-content/uploads/2019/08/NATIONAL-POLICY-ON-ICT-IN-EDUCATION-2019.pdf

Chapter 19

Continental Management of IT

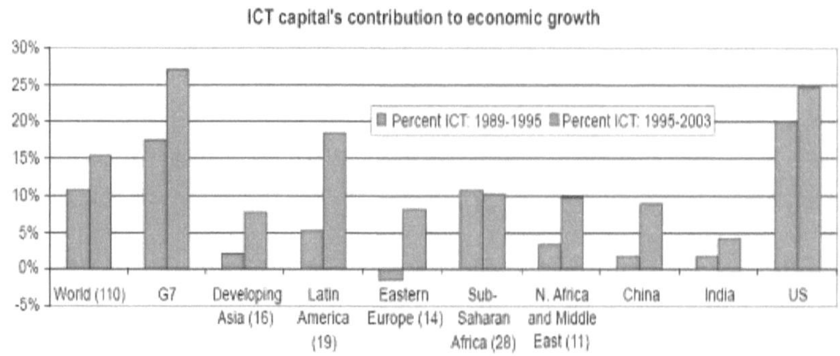

Source: ITU, *World Telecommunication/ICT Development Report 2006*

Judging from the role IT has played in the growth of developed economies, it is pretty clear that IT holds one of the developmental keys for Africa. In other words, IT is a critical factor for development in productivity and businesses, smooth running of government, health, agriculture, education, and others.

IT is a pivotal factor in Africa's economic development, making information easy to get. In our modern times, information is power and wealth. One needs to know what information is required, where it is needed, how to get it, and to whom to sell it to amass wealth. In line with this statement, UNESCO (2005) asserts that societies with thriving intellectual capital are institutions guided by the "capabilities to identify, produce, process, transform, disseminate and use the information to build and apply knowledge for human development."

The availability of the Internet in Africa has made information sharing easier. You get and share information on Facebook, LinkedIn, Instagram, and online newspapers. With the help of the Internet in Africa, communication is straightforward. For instance, African researchers no longer need to go to libraries to get data. Google Scholar has a vast database of scholarly works required for research. Even if you cannot find what you are looking for on Google, you can quickly recruit research assistants from any part of the world online.

Just as information is now easier to access on the continent, education is also becoming easier to acquire. Traditional modes of learning no longer restrict IT-inquisitive Africans. Many Africans today are using the Internet to acquire skills online. There are numerous benefits to undertaking learning online. It is flexible, cheap, and IT-relevant. Another good detail is that most online learning platforms are affiliated with prestigious international institutions and universities. For instance, EdX is an online open course that Harvard and MIT created in 2012. Real college and university instructors and lecturers teach classes on the platform.

Moreover, IT has proven to be a factor in African economic development by providing African youths with a platform to be innovative and creative. Many opportunities accompany IT existence in Africa. Such options include software development, web development, app development for mobile phones, computer programming, cyber security, etc. Though unemployment problems are not yet solved in Africa, the presence of IT opportunities helps reduce its rate. This is because many African youths have acquired considerable IT skills that guarantee employment in the international job market. So, many African youths are employed because of their IT skills; they work at national, regional, continental, and international jobs.

Factors Affecting IT Development in Africa

Numerous constraints are responsible for the drawback and sluggish advancement in IT development in Africa (Dwomoh-Tweneboah, 2008). To compete and be parallel with countries in developed continents, we need to highlight the loopholes responsible for the underdevelopment of IT in Africa.

Low-Level Infrastructure: One of the significant drawbacks to ICT advancement in Africa is a low level of technological infrastructure. As necessary as information is to people, so are the equipment and tools used in providing this information and ensuring the smooth running and spread of its access across Africa. Despite the strides being made by Africa in the increase of mobile adoption and use in Africa, as pointed out in some previous chapters, many studies show that telecommunication diffusion in Africa is the weakest in the world, with the least teledensity (Kelly, 2005; Mureithi, 2017; ITU, 2021). Hence, people have limited access to information. This is one of the reasons why many Africans still depend on public networks, such as cybercafés to browse the Internet, send emails, undertake research, and even communicate with friends and family online. It is essential to mention that the low density of IT infrastructures is affecting other sectors in Africa, for instance, education, where there is limited access to information sources and content, which are highly essential in the research, teaching, and learning process.

Government Regulations and Monopoly. Strict policies and regulations implemented by government agencies are part of the drawbacks putting IT at a disadvantage in many African countries. A typical example is the imposition of high tariffs and the monopoly over IT operations in African countries. Due to this, the exposure of information users to technology operations is limited.

High Cost of Bandwidth. In 2020, the median price of "a gig of mobile data in Africa was more than $5 compared to almost $7 in 2018, a drop of 28%," says A4AI (Benhaddou, 2021). In Europe, "the price was $3.5 per gig in 2020 ($4.1 in 2018)." This results from the high cost of international connection to global telecommunication network sources. The bandwidth cost is always unimaginably higher compared to the world's developed economies (Benhaddou, 2021). Consequently, this high cost affects individuals, academic libraries, research centres, and private information providers. This results from little or no optic fibre sources for local and international traffic or communication. So many African countries depend on satellite technology for connectivity and Internet access.

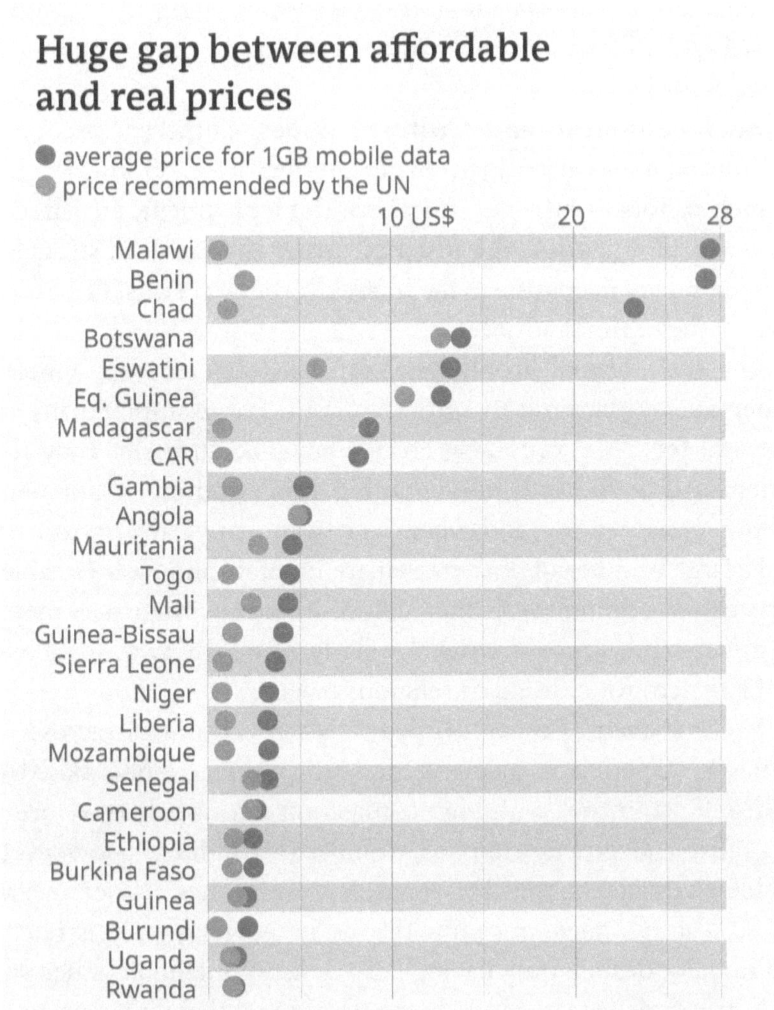

Figure 34: Huge gap between affordable and real prices.
Source: CableUK Research Unit (2020), DW analysis, World Bank, adapted from github.com/dw-data/mobile-internet

Figure 34 illustrates the high cost of bandwidth in Africa.

Low Literacy Levels. Among all information users worldwide, it has been estimated that African users have the lowest literacy level in IT and its infrastructures. Unlike other continents, information users in Africa do not have adequate access to education and training that enable them to operate and navigate the world of IT without obstacles. IT illiteracy affects many Africans in the understanding and operation of IT.

Socioeconomic and Cultural Issues. Some socioeconomic and cultural issues affect IT in Africa. At the socioeconomic level, we see such factors as income, education, unemployment, etc, affecting the use of IT in Africa. For instance, many Africans still struggle to meet their daily needs due to the high poverty rate. Thus, IT becomes a luxury they cannot afford.

On the cultural front, language barriers, religion, customs, gender, and so on, limit the progress of IT. For example, many girls in school feel that computers are for boys. So they shy away from participating in the study of computing, programming, maintenance, computer engineering, and cybersecurity. In many parts of Africa, we have a large number of citizens that are illiterate and so are unable to make use of computers as they communicate in languages they do not understand, speak or write. Similarly, some shy away from using any IT system for cultural or religious reasons.

Inconsistent Power Supply. Many countries across the continent experience inconsistency in power supply. Based on **Figure 35**, countries with high connection rates, such as Mauritius, have reliable electricity supplies, whereas Burundi has a low level of reliable electricity supply. For example, Nigeria has 96% connection and 90% access rates, but only 18% of its power supply is reliable. Additionally, despite only having 23% of its population connected to an electric grid, Tanzania has a high rate of 54% reliability in providing electricity. There are similarities in the continent's problems, but this challenge affects the functionality of IT. IT requires the backup and use of a power supply, and inconsistent power supply significantly affects connectivity across the continent.

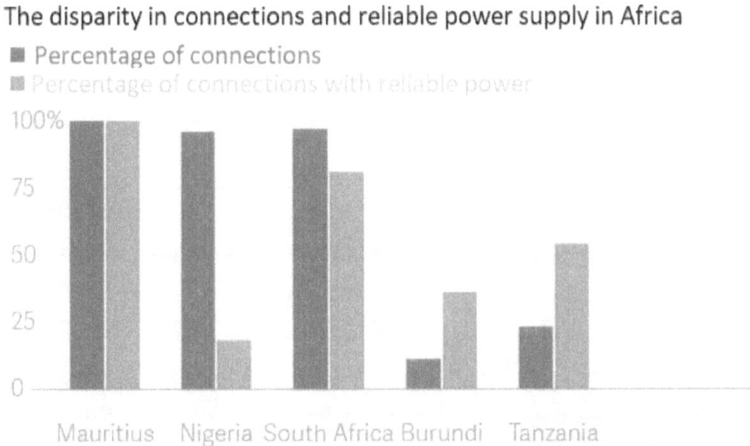

Figure 35: Disparity in power supply in Africa.
Source: ATLAS, Afrobarometer, adapted from *https://www.weforum.org/agenda/2016/04/only-40-of-africans-enjoy-a-reliable-electricity-supply-report-finds/.*

How to Address IT Challenges in Africa

The previous section has shown us some factors affecting IT development in Africa. The steps below are some of the ways to tackle them.

1. Establish a Key Continental IT Programme for Leapfrogging Development

Each continent has peculiar challenges and needs as far as IT is concerned. However, due to colonialism, imperialism, globalisation, and other factors, the African continent has based its IT programmes and development on the models of developed economies. This acquiescence is not healthy for the continent. Therefore, the continent must establish continental IT programmes to help it leapfrog to rapid growth.

Leapfrogging in this context means rapidly moving from the current unsatisfactory IT stage to the enviable level of IT development obtainable in developed economies. On the other hand, continental IT programmes are programmes capable of solving the particular challenges facing a specific continent, in this case, Africa. Such programmes should assist African countries in achieving the same IT feats as developed countries (Enakrire & Onyenania, 2007). The following four(4) paragraphs discuss some IT programmes that can help Africa leapfrog to the desired IT development.

As it is pretty clear to Africa that it lags in IT development, just like it lags in other areas, the first thing is to develop continental IT policies and initiatives that favour IT growth across the continent. To do this, there should be extensive continental consultation with the major stakeholders in the IT sector: IT experts, technicians, lecturers, students, etc. These stakeholders should be able to point out the fundamental challenges to address. Apart from formulating IT policies at national levels, African states and governments should join forces at the continental level. However, this task should not be left to the states and governments alone; continental partners for the progress of IT are vital. For instance, in October 2007, African Development Ban introduced *Connect Africa Initiative*, which aims to mobilise the human, financial, and technical resources that can close the ICT gaps on the continent (African Development Bank, 2007).

More so, we should have many IT programmes focusing on science, technology, engineering, and mathematics (STEM) education. Because of colonial legacies, the educational systems of African countries do not have a solid STEM background. Without strong STEM, it would be hard for African countries to achieve IT-driven development at the continental level. Therefore, a programme based on STEM education should be encouraged. Achieving IT programmes through STEM will involve using ICT to improve the quality of teaching and learning by equipping teachers with intensive IT skills, ensuring the availability of IT equipment and applications in schools, building living labs (LLs), and innovation spaces and centres (InnoSpace). To make these possible, the educational curricula of institutions should be overhauled and changed to make IT development possible.

The IT programmes should encourage IT research and development (R&D). Africa's ability to leapfrog IT development is a mirage without strong R&D on the continent. With functional R&D across the continent, it will be easy for IT researchers to conduct researches that address the benefits, trends, challenges, and solutions of IT in Africa. One way of achieving these is to establish IT R&D centres across the continent. Another way could be to partner with other bodies interested in IT research in Africa, for example, the International Journal of ICT Research and Development in Africa (IJICTRDA), whose mission is to provide intensive research, advanced analytical methods, techniques, e-innovations, and development policies on the continent and beyond.

Continental IT startups are significant for leapfrogging IT development. Therefore, the continent should formulate and implement continental IT programmes for startups to proffer homemade solutions to tackle the challenges confronting IT development. The continent should learn from what some groups and big tech companies are doing in Africa. For instance, Google for Startups Accelerator Africa was launched in 2017 for 17 African countries to help African startups to scale through.

2. Develop Strategic Plans

Programmes without plans cannot succeed. Therefore, IT development across the continent will need an IT strategy plan. The strategic plan should include the strategies, how to implement them and what is required to implement them, and the execution and assessment.

In terms of the strategies to use, the continent should look at the challenges confronting it in IT. Doing that will provide more precise directions to follow. Also, the trends identified should guide the aim and objectives or mission of Africa's IT strategy plan.

Besides, choosing strategic plans should involve and detail all the necessary tools, in which case, SWOT analysis becomes relevant. In other words, the continent's strengths, weaknesses, opportunities, and threats to IT development should be evaluated. For instance, no IT strategy plan for the continent can be efficient without considering the infrastructural threats that invalidate them.

To develop Africa's IT strategic plans, action is tantamount; in other words, whatever IT plans are chosen should be executed appropriately. For instance, IT strategic plans are not limited to installing telephone lines and computer systems to the Internet. All the relevant tools should be provided for execution, and one of the ways of doing that is through wide bandwidth.

Re-Architect Africa with Very Wide Bandwidth

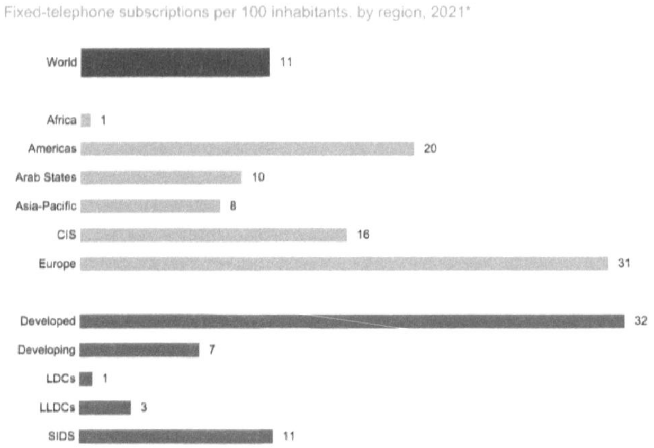

Figure 36: Fixed telephone subscribers per 100 inhabitants by region.
Source: ITU Estimate

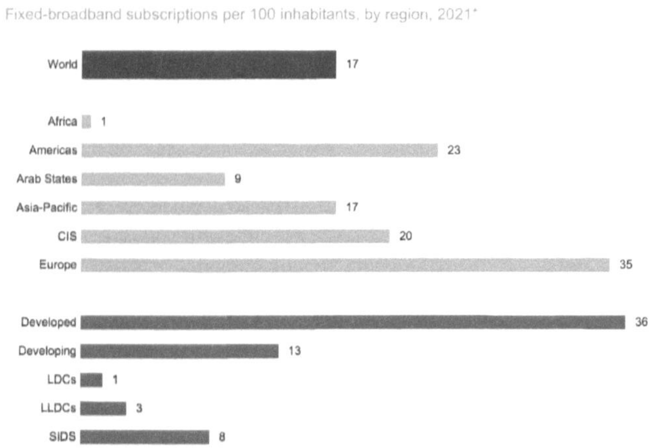

Figure 37: Fixed-broadband subscriptions per 100 inhabitants.
Source: ITU estimate

Figure 37 shows that fixed broadband subscriptions are steadily increasing, with a global average of 17 subscriptions per 100 inhabitants assumed in 2021. Despite double-digit growth, fixed broadband remains a luxury for a select few in LDCs, with only 1.4 subscriptions per 100 inhabitants. Since 2017, the number of fixed broadband subscriptions has exceeded that of fixed telephony globally (see **Figure 36**). Asia-Pacific recorded the highest growth, at 10.5%, followed by the Commonwealth of Independent States (7.0%) and Africa (6.7%), which had the highest growth over the last three years.

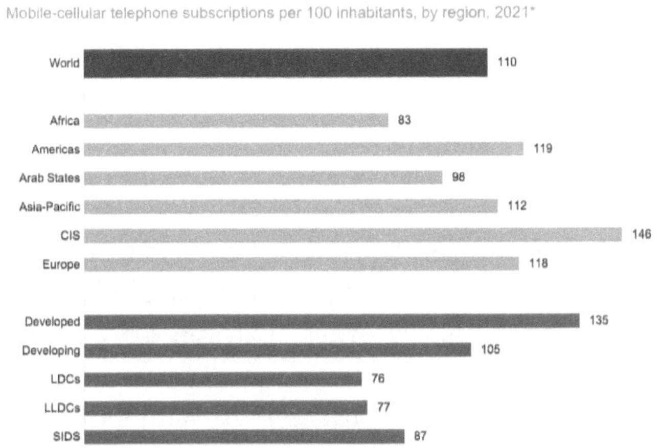

Figure 38: Mobile-cellular telephone subscriptions per 100 inhabitants.
Source: ITU Estimate

Figure 38 shows that the penetration of mobile cellular subscriptions increased once more in 2021, reaching a record 110 subscriptions per 100 inhabitants after a slight decline in 2020. The same pattern was seen for mobile broadband subscriptions (3G or better), which reached 83 subscriptions per 100 people. Developing nations in the Asia-Pacific and Americas regions were the main drivers of this increase. Contrarily, mobile cellular subscriptions in Africa fell in 2021 after rising in 2020.

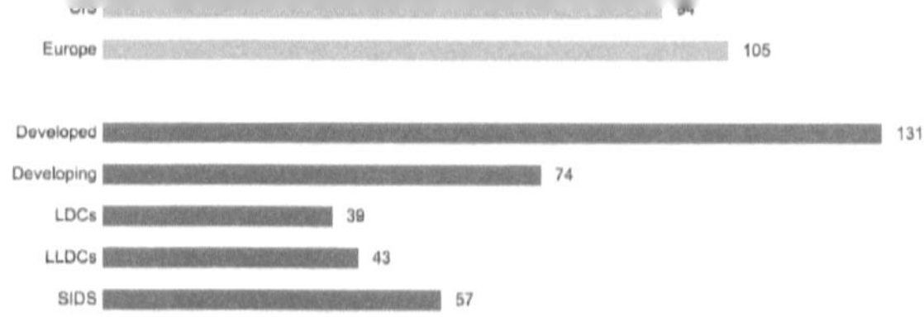

Figure 39: Active mobile-broadband subscriptions per 100 inhabitants.
Source: ITU estimate

Figure 39 shows that Africa needs extensive bandwidth to accelerate IT development. Technology is like a living organism; its growth and development require nurturing. Therefore, Africa needs to re-architect its IT programmes with vast bandwidth. For instance, in developed economies, the 4G network is no longer in vogue; it is now a 5G network. However, here in Africa, we have only begun to embrace 5G. No wonder the African Development Bank (AfDB) argues that "the scarcity of cross-border backhaul links is one of the key broadband access gaps in Africa" (African Development Bank Group, 2019).

Why does Africa need a vast bandwidth? With the increased data demand for connectivity and information, smartphones, computer systems, tablets, and other Internet-accessible devices need high-speed network capacity to ensure every individual can access and navigate the Internet without experiencing connection glitches.

How, then, can Africa achieve vast bandwidth? According to the World Bank, Africa can achieve bandwidth by investing $100 billion. This will enable it to "achieve universal, affordable, and good quality internet access by 2030." It is one of the surest ways of closing the Internet access gap on the continent (World Bank Group, 2019). This investment will enable Africa to have substantial finance for optic fibre networks, which can help Internet Service Providers (ISPs) and mobile users to have faster and better broadband. The investment will also help the continent to address power challenges, as we know that vast bandwidth cannot be possible without stable electricity. The investment can also help the continent to upgrade to 5G networks across the continent.

Open Source Software Development and Usage in Africa

Open Source Software (OSS) refers to publicly available software systems that users can use as a result of the developer granting users free access to make use of the software as the user pleases, not only the right to utilise the software as necessary but also carry out modifications and distribution of the modified version of the software the developer made. The introduction and application of Open Source Software (OSS) have contributed significantly to the development of IT in all parts of the world. One of the remarkable features of the software is its ability to bridge the technological gap at a relatively affordable cost.

In terms of the development of OSS in Africa, end users mainly depend on foreign ones. However, that does not mean Africans are not making efforts to develop OSS. Many African developers and stakeholders have taken bold steps to ensure the development of OSS. For instance, there are some OSS built in Africa; some of them include Mbuni (Uganda), WooCommerce (South Africa), FrontlineSMS (Kenya), RootiO Radio (Uganda), and Ushahidi (Kenya).

OSS has many benefits for Africans. For instance, OSS has a great potential to save costs and reduce business costs. Its massive adoption will enhance the training of software engineers because they will benefit from the expertise of the OSS community. The development and usage of OSS in Africa can significantly influence economic outcomes; as we know, OSS can be applied in governance, commerce, health, education, and agriculture, among others.

However, there are many challenges affecting the development and usage of OSS. They include lack of OSS awareness, lack of OSS policies, government/R&D support, language barrier, poor Internet connectivity, and lack of technical competence required to use OSS. For instance, African governments have not provided the requisite infrastructures, such as stable electricity, high Internet connectivity, wide bandwidth, and affordable data, making OSS development and usage easier.

Tackling the challenges affecting the development and usage of OSS in Africa would need a collective effort. In other words, governments, educational institutions, and private and public organisations have roles to play. For instance, African states and governments must develop policy guidelines to guide the OSS industry. The OSS policies should involve accurate implementation, usage, and maintenance. On the part of educational institutions, relevant departments in IT should organise OSS workshops, seminars, conferences, etc.

Building Infrastructures for the Development of Information Technologies in African Countries

In pursuit of the advancement and rapid development of information technology in African countries, the building and availability of infrastructure are imperative. The IT infrastructure is equally as critical as the information itself. They help facilitate, improve and organise the free and smooth running of information without issues. The required IT infrastructure will put African countries at shoulder level with countries in developed economies.

Policy formulation is the first step toward building infrastructures for IT development in Africa. No meaningful action happens without well-planned policies. Therefore, African countries should formulate programmes to ensure that IT infrastructures are available on the continent. To develop IT infrastructural policies, relevant stakeholders in the IT industry and the citizens should be consulted. For instance, the R&D of African countries should be tasked to research how to make IT infrastructures feasible.

Besides, there should be an appropriate budget and funding for the IT programmes. To achieve this, major stakeholders and government agencies would play a significant role in providing funds or capital for the purchase, instalment, and ignition of these infrastructures. On the part of African governments, for instance, they should make adequate budgetary provisions for IT infrastructures. As for both private and public African partners, they can sponsor IT infrastructural programmes.

The last step in the building of IT infrastructures is execution. This involves using the funds allocated for IT development to implement IT policies and to purchase and install the necessary infrastructures. However, investing in human skills is essential for such infrastructures to thrive and be of any function. So, there should be regular education and training of human resources to help advance knowledge and ensure the ability to operate and navigate these infrastructures to provide glitch-free results.

Setting Guidelines for Equipment Manufacturers, Distributors, Service Providers, and End Users of IT in Africa

Specific guidelines and policies should be implemented to ensure the safe and smooth running of IT infrastructure development, distribution, service provision, and usage in Africa. The guidelines may apply to infrastructure supporting storage, network devices, and IT services. For African states and governments to set proper guidelines for IT equipment manufacturing, distribution, and usage, they should consider the following:

- Need
- Standard
- Monitoring
- Education and enlightenment of end users

In areas of need, African states and governments should set guidelines and policies to solve peculiar IT challenges on the continent. The procedures should be based on the IT needs of Africans. For instance, African governments should mandate local and multinational IT equipment manufacturers, distributors, and service providers operating on the continent to have a detailed local content development plan for job creation for Africans, human capital development, and the use of indigenous African IT.

To ensure the standard of IT on the continent, African governments should set up and empower relevant agencies to issue licences and certifications only to qualified IT manufacturers, distributors, and service providers. The licence and certificates should be subject to renewal. The IT guidelines should also ensure that equipment manufacturers, distributors, and service providers design and develop IT products that support African languages and local use. It is essential to prioritise the benefits and relevancy of IT equipment to ensure that they meet the required need of users. In the education sector, for instance, priorities should be given to developing open-source software that allows students access to information resources that would be helpful to the learning process.

Besides, it will be tough for IT manufacturers, distributors, and service providers to adhere to standards without serious monitoring. Government and public agencies should be empowered to monitor local and multinational IT devices and equipment to ensure they conform to the continent's standards. For instance, African states and governments must watch the IT manufacturers, distributors, and service providers to ensure they do not charge exorbitant tariffs and prices. African governments can create apps and software for monitoring them.

As all these measures benefit African users, governments should also set guidelines for educating people about their IT policies and programmes. This will help users to know when and how to hold IT manufacturers, distributors, and service providers accountable. For instance, carrying African end users along will assure users the protection of their personal information. This is only possible when African end users are enlightened about what constitutes breaches of their data security and safety.

The Acquisition of Information Technology in Africa

Acquiring IT means purchasing IT machinery or equipment and hiring and contracting IT specialists. Africa must develop quality, relevant, and specific IT software and hardware to leapfrog to enviable IT development. The IT acquisition could be made internally or externally. On an internal scale, the acquisition can be made by patronising local IT specialists and equipment. On the other hand, external IT acquisition involves making use of IT experts and services from different continents or multinational experts and equipment resident in Africa.

IT acquisition is a process. The first process is planning. African governments must develop tactical and strategic plans that ensure long-term IT infrastructure to support business activities and other essential sectors. The IT acquisition plans should have both short-term and long-term dimensions to be responsive to emerging opportunities as the IT industry continually evolves and improves to provide more essential services. More importantly, when acquiring information technology, it is imperative to align the beneficial priorities of these technologies to the continent's IT goals.

The following process of IT acquisition is the implementation and execution of the laid-out plans and policies. This stage involves setting up IT acquisition teams, funding and purchasing IT equipment, or contracting IT specialists. For instance, IT acquisition agencies must thoroughly analyse these technologies' benefits, costs, risks of infrastructure obsolescence, feasibility, and maintenance processes.

Lastly, the plans should be reviewed from time to time. There should be an assessment of the IT plans, procurement, implementation of IT, and their impacts. The evaluation will determine the progress and setbacks recorded in the IT acquisition.

Sectoral Areas of Application of Information Technology

IT has significant effects on all sectors of an economy. It is applied or used in health, education, business, finance, agriculture, government, transportation, telecommunication, and many more.

Electronic Commerce: Information technology can be used for commercial processes – unlike traditional buying and selling in an open market. With the advent of information technology and allied software, consumers can buy goods and services online. This is for the benefit of both buyers and sellers.

E-Government: IT is used in governance across Africa. Governments apply IT in administration, legislation, judiciary processes, elections, etc. With the help of IT, management is becoming easier on the continent. For example, government conferences can be held online.

Electronic learning: Over the years, information technology has contributed mainly to education all over the world. With the use of IT, information contents can be disseminated and made available to a large number of people remotely. All that is needed is to use computer systems and mobile data information. Recently, online classes have been introduced that allow students and learners to study and execute assignments and examinations via the Internet. Unarguably, this has greatly improved the quality of education across Africa.

E-Health Care: This is one of the exciting uses of information technology. With the presence of information technology in our present day, healthcare services can be provided virtually by specialists in various medical departments. Today, virtual medical consultants are readily available to provide services to patients. Information users now have access to health-related data on the web, which enlightens them in their knowledge about specific health issues and diseases. For example, since 2021, after the invasion of the deadly COVID-19, the World Health Organisation has taken necessary procedures to ensure that valid and accurate information is available to people online on how to curb the spread of the disease or manage it.

The Development of Human Resources

Human resource is integral to any striving organisation hoping for great productivity. This is because human resource development is the framework for helping personnel develop and improve their skills, knowledge, and basic abilities, which in the long run, would reflect on the organisation's productivity.

Human resources are essential for IT development. However, they are not yet in the required number to bring about the required IT development on the continent. In other words, there is a dearth of IT experts in Africa. As vital as IT is in Africa, it is evident that these technologies would not operate themselves. There is a need to develop human resources in the operation process of information technology to enable user satisfaction and meet the IT plans for the continent.

One way of ensuring the sufficient availability of human resources in Africa is re-training and training IT personnel. African governments should re-train the existing IT experts. They should also make provisions for training new experts. This will ensure that the continent will get sufficient IT human resources to bring about meaningful development in the IT sector. It will also result in increased productivity, low cost of employing experts, regular maintenance of information systems, support, and workforce collaboration. Also, IT evolves at rapid intervals. So, regular seminars and training on the use and application of newly invented technologies should be organised to help facilitate their benefit to humans.

Another way to ensure the availability of IT human resources is through education. Education curricula in African countries should be overhauled to accommodate IT skills. This will encourage African youths to take courses in IT, and when they graduate, they will increase the IT workforce on the continent.

Regulation, Deregulation and Management of Information Technology in Africa

For the success of IT in Africa, governments and relevant agencies should regulate, deregulate and manage IT. Regulation means making rules and regulations that IT firms and users are expected to observe, while deregulation means removing or relaxing government *regulations* on the IT industry. Regulation/deregulation and management of IT are critical to the development of IT in Africa. They ensure that IT industries operate businesses more smoothly and satisfy the IT needs of Africans.

The regulation also ensures the protection of individual privacy, regulatory cost of bandwidth, and subscription tariff. It will also eliminate the sector's monopoly by influential local and multinational investors. The IT regulatory policies should not be too rigid; they should be subject to reforms. One of the objectives of reforms is to ensure deregulation where and when necessary.

Deregulation aims to eliminate rigid and non-functional government laws that do not allow IT progress and development. Deregulation will promote overall IT operations, usage, and functionality among the people in the country. However, deregulation should not be too flexible; otherwise, it will lead to many challenges that can stifle the development of IT in Africa.

Overall, Africa's IT regulation, deregulation, and management are crucial for the continent's economic, social, and political development.

References

African Development Bank (2007, October 29). AfDB and ITU join forces to Interconnect African Capitals and Major Cities. Retrieved May 12, 2022, from https://www.afdb.org/en/news-and-events/afdb-and-itu-join-forces-to-interconnect-african-capitals-and-major-cities-2595

African Development Bank Group. (2019, April 23). *ICT Initiatives.* African Development Bank. https://www.afdb.org/en/topics-and-sectors/

sectors/information-communication-technology/ict-initiatives

Benhaddou, S. (2021, July 13). *Africa: Which countries charge the most for internet data?* The Africa Report. https://www.theafricareport.com/107259/africa-which-countries-charge-the-most-for-internet-data/

Dwomoh-Tweneboah, M. (2008). Information Technology for Africa. *Revista de Educação do Cogeime*. 17. 165-179.

Enakrire, R. & Onyenania, G. (2007). Factors affecting the development of information infrastructure in Africa. *Library Hi Tech News*. 24. 15-20. 10.1108/07419050710751643.

ITU. (2021). *Digital trends in Africa 2021: Information and communication technology trends and developments in the Africa region, 2017-2020*. International Telecommunication Union Publications; ITU Publications Africa. https://www.itu.int/dms_pub/itu-d/opb/ind/D-IND-DIG_TRENDS_AFR.01-2021-PDF-E.pdf

Kelly, T. (2005). Changing ICT Rankings of African Nations. *South African Journal of Information and Communication*, *5*, 40–48.

Mureithi, M. (2017). The Internet Journey for Kenya: The Interplay of Disruptive Innovation and Entrepreneurship in Fueling Rapid Growth. In: Ndemo, B., Weiss, T. (eds) Digital Kenya. Palgrave Studies of Entrepreneurship in Africa. Palgrave Macmillan, London. https://doi.org/10.1057/978-1-137-57878-5_2

UNESCO. (2005). *Investing in Creativity - Diversity of cultural expressions*. UNESCO. Retrieved May 12, 2022, from https://en.unesco.org/creativity/sites/creativity/files/info-kit_brochure-final-en-web.pdf

World Bank Group. (2019, October 17). *Achieving Broadband Access for All in Africa Comes with a $100 Billion Price Tag*. The World Bank. https://www.worldbank.org/en/news/press-release/2019/10/17/achieving-broadband-access-for-all-in-africa-comes-with-a-100-billion-price-tag

Chapter 20

IT and Youths in Africa

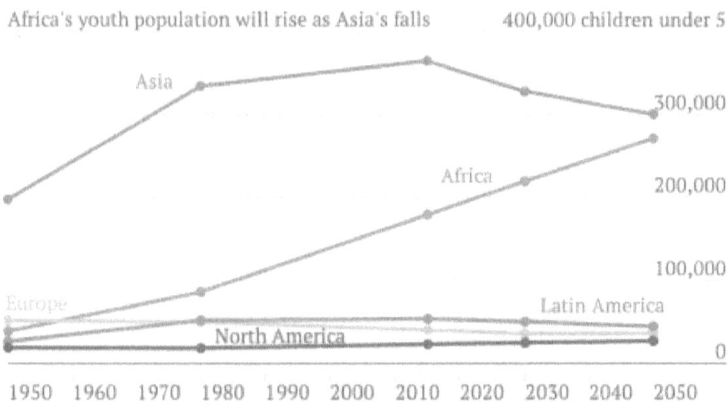

According to the United Nations (n.d.), youths are 15-24 years old. Though the UN puts the age range of youth from 15-24, it acknowledges that the content could vary in the member states. Therefore, the age bracket of youth in Africa differs from that of the UN. For instance, in the African Youth Charter, youth ranges from 15-35, in the National Youth Policy of Nigeria (2019) from 18-29, in the National Youth Policy of Ghana (2010) from 15-35, etc. The age-range difference in Africa reflects political and cultural realities.

On the other hand, demography refers to the study of human populations. The UN identifies indexes of demography to include size, growth rates, composition, etc. So, the demography of youths in Africa refers to the population of youths based on the age brackets identified by the UN or the African member states.

Among the comity of continents, Africa has the highest number of youths, with almost 60% of the continent's population under 25, according to the UN. Interestingly enough, Africa's young population has been on the increase. From UN statistics, in 2015, it was 226 million youth; by 2030, it is projected to increase by 42%, and by 2055 (United Nations, 2015), it will be more than double the current levels.

Many factors have been identified as the reasons behind Africa's highest number of youths. One of them is polygamous marriage. From the old days, polygamy was common among African men. Although, polygamy is gradually waning, it is still practised in some parts of the continent. One result of polygamy is many children. In some families, a man can have ten children and above.

Another factor is early marriage. For a long time, early marriage has been practised in Africa. Even in these modern days, early marriage is still practised in many parts of Africa, with some girls being married off below the age of 18. Early marriage implies that the young couple gives birth early. In some families, the early marriage tradition continues (Kariba, 2020).

Consequently, having the youngest population in the world has both positive and negative implications for the continent. On the positive side, the young population provides agile, active, and constant labour (Population Reference Bureau, 2019). The surplus manpower positively impacts African economies; it can drive economic development if properly managed by governments. This is because youth is a stage where creativity and innovation reign supreme.

Presently, many African youths are facing numerous harrowing experiences and prospects. In the first place, African governments have not been able to provide jobs for the teeming youth population. According to the African Development Bank Group (2020), 11 million young Africans join the labour market each year. Still, the continent generates only 3.7 million jobs annually, leaving most youth demography with fewer options. The prospects of the unemployed youths are to look for employment in the informal sectors, acquire additional skills, become self-employed, and migrate out of Africa.

At times however, some of the youths choose extreme options. Out of frustration and despair, some unemployed youths resort to drug pushing, armed robbery, kidnapping, prostitution, terrorism, etc. For instance, "the World Bank estimates that 40% of people who join rebel movements are motivated by a lack of economic opportunities" (Canagarajah, 2014). These societal ills create chaos on the continent, making it hard for sustainable development.

Opportunities in IT for African Youths

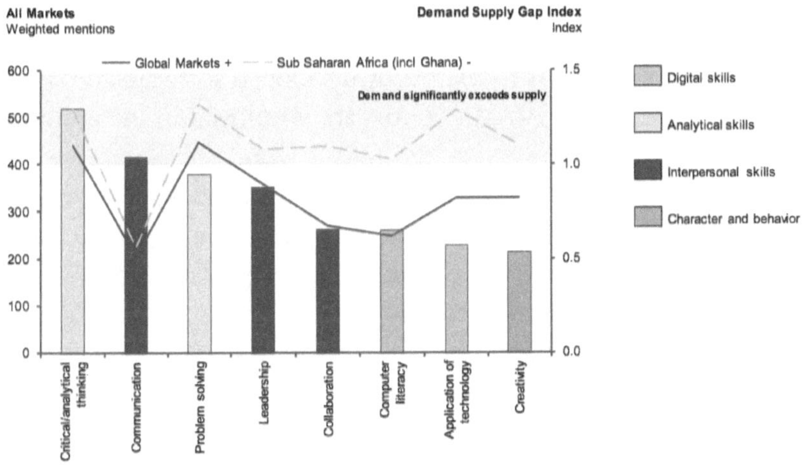

Figure 40: Supply and Demand for Most Important Workforce Skills
Source: *World Economic Forum*

The preceding section has shown the ugly side of the high population number of youths in Africa. Therefore, it is imperative that African leaders bear that in mind and ensure that their youth bulge becomes a net positive for their countries, rather than a burden. By so doing, they can make the best use of IT to engender development for their respective countries (MINDS Connect, 2018; Nwokoma, 2021).The rise of IT and the effects of the COVID-19 pandemic have made it possible for people, including African youths, to work from anywhere in the world as long as they have the requisite IT skills.

The IT opportunities available for African youths are listed and discussed below:

1. Software Development: Software development is an IT job that creates and builds software for systems. In other words, it involves specifying requirements, designing applications and writing codes for systems and mobiles. Software developers or engineers write codes and software programmes to perform chosen tasks. They write the code based on instructions and guidelines they provide through a programming system. Many African youths have both formal and informal job opportunities in software development. For instance, opportunities exist at the state, continental, and international levels.

2. Computer Programming: Computer programming is a subset of software development. It involves writing the code required to get the system to work. Additionally, it is the job of computer programmers to test newly developed software and update older software. Many African youths are employed in the computer programming sector.

At the international level, Big Techs like Facebook, Google, IBM, etc, have employed many African youths. For instance, on 30 August 2016, the founder and CEO of Facebook, Mark Zuckerberg, visited Lagos, Nigeria. In Yaba, he stopped at Co-Creation Hub (CcHUB) and Andela. He held a town hall meeting with IT entrepreneurs and developers. Moreover, he invested $24 million in Andela through the Chan Zuckerberg Initiative in 2016 (Konrad, 2016).

3. Cyber Security: In IT, cyber security experts are responsible for protecting the security of computer systems and networks. This profession needs sound IT knowledge because security is integral to a modern computer system. Therefore, firms, businesses, corporations, and governments have employed many African youths in cyber security.

4. Data Analysis: Data analysis involves gathering, evaluating, and gaining hidden data insights. So, a data analyst is needed in every organisation to help it analyse and gain insights into the data they use. As organisations, firms, and governments deal with extensive data, the job of a data analyst becomes imperative. The analyst collects, analyses, and interprets the data. As data is essential for the progress of any business, organisation, or government, there are many job opportunities for African youths in that IT domain.

5. Web Design/Development: This is the branch of IT that involves designing, building, or developing websites for blogs, individual businesses, organisations, and government ministries and departments. With the surge of the possibilities of the Internet, the demand for web builders has increased. Many African youths have been formally or informally employed in the web development sector at national, continental, and international levels.

6. Mobile Applications Development: The demand for mobile applications (apps) has increased with mobile phones and connectivity. By mobile apps, we mean the application software designed to be used on mobile devices, such as smartphones or tablet computers. Mobile applications provide mobile device users with similar services to those accessed on their PCs. Many African tech-savvy youths have developed mobile apps to solve health, education, financial inclusion, agriculture, and other socioeconomic challenges. Those mobile apps include M-Pesa, Paystack, iCow, MedAfrica, etc. For instance, Paystack, a fintech payment app, was developed by two Nigerian youths, Ezra Olubi and Shola Akinlade, in 2016. The app was acquired by a giant Irish-American financial company called Stripe. This app has given many jobs to numerous African youths.

7. Social Media Management: Social media is an umbrella term for interactive technologies that share information, content, and ideas online. Some popular social media include Facebook, Twitter, TikTok, Instagram, Wimi, YouTube, Quora, LinkedIn, Trello, Mighty Networks, and many more. Social media management involves creating and sharing content/information with a social media audience. Because high-profile people, businesses, and governments have social media profiles and pages, they recognise the necessity for social media managers. The social media managers are responsible for developing content and strategies that align with the followers and audience of a social media account profile. Many African youths have jobs in the social media management domain.

8. Digital Marketing: With the emergence of the Internet, virtually everything can be bought and sold online. As a result, digital marketing is the branch of marketing that involves utilising computers and mobile phones to advertise the products and services of individual enterprises and companies. For instance, Jumia, Konga, Kilimall, and Takealot, are African e-commerce websites that give many youths jobs.

9. Other Areas of Specialisation: In addition to the areas above, African youths can specialise in many other areas. For example, they can specialise as hardware operators, network engineers or administrators, robotics engineers, AI developers, machine learning specialists, blockchain developers, and IOT/IOE specialists.

Challenges to IT Opportunities for the Youths in Africa

It is agreed that there are many job opportunities for African youths, but challenges hinder the realisation of opportunities. It is necessary to identify and discuss these challenges so that measures to address them can be proffered.

i. **Power Challenges.** IT opportunities thrive well in countries with stable and constant power supply. Unfortunately, the power supply is a severe problem for various governments and administrations on the African continent. The lack of stable and constant electricity makes it hard for many African youths in cities and rural areas to embrace IT opportunities as satisfying and readily accessible as possible. For instance, constant electricity is needed to charge their laptops, phones, and other devices that can keep them on the Internet to get IT job opportunities. With poor power supply and the high cost of standby generators and solar panels, access to IT opportunities for African youths is limited.

ii. **Weak Internet Connection.** For IT opportunities to be possible for the majority of African youths, a strong Internet connection is sacrosanct (Kalan, 2013). Sadly, there are weak Internet connections in most parts of the continent. In worst cases, most of the villages and remotest parts of Africa do not have mobile network coverage, bringing about an abysmal digital divide. The youths in such areas are cut off from all the opportunities IT offers to the digital world.

iii. **Government Policies.** Government policies determine the progress or failure of IT in countries. In other words, a continent whose countries make IT-friendly policies will make IT opportunities possible for their citizens and vice versa. Unfortunately, most government programmes and policies in Africa are not favourable for IT opportunities to thrive. For instance, Nigeria banned the trading of cryptocurrency. This policy deprives the Nigerian youths of the opportunities offered by the new financial technology.

iv. **Ignorance and Lack of Access to IT Information.** Most African youths have mobile phones. Still, many of them do not know the enormous IT opportunities available on the Internet. Many African youths still see mobile phones as devices for calls and chats alone. They do not know most of the other things, for example, Google, YouTube and Internet courses they can use their mobile phones for. For instance, many African youths are still looking for someone to teach them skills without knowing they can learn any skill on YouTube, Udemy, Coursera, etc. Worst still, African youths in rural areas may not have heard of some of these vital mobile apps.

v. **High Cost of Data.** To surf the Internet, mobile data is needed. Sadly, most African countries have a high cost of data. Mobile data service providers charge high. Therefore, it is hard for poor African youths to take advantage of IT opportunities. African youths who are unemployed or dependent on their parents cannot cope with the high cost of data.

Education, the Digital Economy, and the Future of Work

The digital age is a disruptive stage of human history. In other words, digital technology has dramatically changed all aspects of human lives, education inclusive. This digital revolution in education implies that traditional ways of learning are rapidly forced to be changed to suit the demands of the digital world (Karuitha, 2020).

The change in education is seen in the rapidly growing use of Internet-based learning. Though many developed countries have been experimenting with virtual learning, the COVID-19 pandemic forced many countries to adopt online classroom experiences. Students have been taking courses online. From how the online learning experience has been going since then, it is evident that virtual learning has come to stay.

Besides, the digitalised world has introduced a lot of new courses. Many curricula have been changed, especially in the world's developed economies. Learners are now trained in techniques like video editing, mobile apps, software development, digital marketing, virtual assistance, and Search Engine Optimisation (SEO).

As a result of the curricular changes in traditional education systems, extraordinary digital classrooms/schools are now in vogue. One can choose to school online; one does not need to go to traditional institutions to be educated or to acquire skills. For instance, you do not need a university education to be a programmer, graphic designer, or digital marketer. You can quickly access these courses on Coursera, Udemy, YouTube, FutureLearn, and other online learning platforms to get any digital skill you want.

Just as the digital revolution has changed education systems, the economy has also changed. We are now in the era of the digital economy. We can see that the 4IR technologies are now replacing traditional modes of production, distribution, consumption, and governance. With the 4IR, we have new industries, jobs, higher GDP growth, and better living standards. For instance, most of the jobs and activities hitherto being done by humans are now automated. AI and process automation are rapidly replacing human efforts in several domains of endeavour. Therefore, human actions are being channelled into other creative and innovative work.

Moreover, the digital economy has made specific traditional skills obsolete. The demand for new skills has emerged; digital and tech skills are now in higher need. In contrast, traditional skills face endangerment. Today we speak of emerging skills such as complex problem-solving, analytical thinking and innovation, active learning and learning strategies, and so on. On the other hand, some skills are now being referred to as declining skills. Some declining skills include technology installation and maintenance, as well as skills requiring manual dexterity, endurance, and precision.

In the market, some jobs are now emerging, while others are declining. For instance, jobs in the blue-collar (e.g., factory workers, truck drivers, and construction workers) and white-collar (for example, clerks, secretaries, bookkeepers, and accountants) categories are declining. On the other hand, some jobs are in higher demand. Some emerging jobs include data analysis, data science, software development, web design, computer programming, AI and machine learning, and so on.

Furthermore, the digital economy has paved the way for the gig economy. It is no more the traditional work style as usual. You can work from home based on the work arrangements that fit your employers. Instead of employing staff, companies and organisations now rely on outsourcing and contract. People can now stay at home and get gigs from different firms.

The current gig economy indicates that the future of work is automation and AI. A significant percentage of human jobs and work would be taken over by machines and AI (Annabel et al., 2020). Job automation has both negative and positive sides. On the negative side, humans who do not have the requisite digital skills will be made redundant, thereby leading to more unemployment. On the positive side, the automation of jobs will improve performance and productivity, reduce drudgery, and give human beings the privilege to focus on other problem-solving activities.

Overall, the digital economy has tremendous effects on education and the future of work. Therefore, individuals, businesses and organisations, educational institutions, and governments are constrained to adapt to the digital economy's demands sooner than expected. For instance, governments around the world need to make policies that favour and promote the digital economy; educational institutions need to change their learning systems to be in pace with the digital economy; employees need digital skills to be relevant in the labour market, and businesses and firms have to make provisions for automation and other digital-friendly work arrangements.

African Youths and the 4IR

Though the older generation of Africans has its role to play in achieving the goals of the 4IR on the continent, the success or failure of the 4IR in Africa is in the hands of the African youths. The African youths are still agile and active, and the 4IR era is taking place in their generation. Therefore, this section raises questions: what are the 4IR opportunities available for African youths? How have the African youths been performing in the 4IR scheme of things? What challenges make being part of the 4IR difficult for African youths?

To answer the first question, we have to look at the job opportunities made possible by the 4IR technologies. The 4IR has many underpinning technologies, including 3D and 4D printing, horizontal and vertical system integration, additive manufacturing, augmented reality, virtual reality, autonomous robots, artificial intelligence, cloud computing, cyber security, etc. All these technologies, in one way or the other, have job opportunities for

African youths.

Moreover, many African youths have taken advantage of the 4IR technologies to carve niches for themselves. They have positioned themselves in these 4IR domains. For instance, many have designed and developed computer programmes, websites, software, mobile apps, etc. Also, some innovative African youths have taken it upon themselves to teach others the 4IR skills they have acquired. For instance, Ghetto Heroes Tech Innovation is a South African non-profit organisation founded by a young South African, Diketso Setho, who taught himself coding and started coding in high school (Hastings-Spaine, 2021). His organisation teaches coding to kids and youth located in disadvantaged communities. A group formed by some youths in Nigeria, Alkebulan Tribe, has created thousands of NFTs (Non-Fungible Tokens) comprising African Digital Arts. This speaks very much to the 4IR era and shows that they are riding its wave.

However, many African youths are still lagging in grabbing the opportunities offered by the 4IR. They lag because of some factors, such as lack of awareness about the possibilities of the 4IR, lack of the requisite 4IR skills and competence, power challenges, government policies, and the high cost of 4IR facilities. Therefore, African governments have roles to play to ensure that most youths take advantage of the 4IR technologies, get employment, and help develop African economies.

Ways African Countries Can Prepare Their Youths for the 4IR

Despite the fact that the African youths are the champions of 4IR success in Africa, the framework for the optimal success of 4IR in Africa has to be streamlined and directed by African governments.

The first step African countries can take in preparing their youths for the 4IR is to make 4IR-compliant policies. The 4IR has advanced the economies of the developed nations because the governments made and keep making policies that favour the emergence and evolution of the 4IR. Therefore, to make considerable progress in the 4IR, African countries and their governments should formulate policies that encourage the growth of the 4IR. They should invest in 4IR Research and Development (R&D). It is not enough to make just policies; implementation is imperative. Changing Africa's education system is a practical demonstration of the implementation.

The initiative by the South African government in that direction is commendable. It has developed "curricula for coding and robotics for grades R to 9." The Minister of Basic Education, Angie Motshekga, announced this during the 2020/21 virtual plenary session on the adjusted budget vote. She said that this would enable South Africa to create sustainable industrialisation to keep pace with developed economies. She added that the curricula would produce learners with the skills needed for future work (South African Government, 2020).

Another way to implement the 4IR policies successfully is through re-orientation. Many African youths do not know the enormous and boundless possibilities in the 4IR economies. Most youth graduates are busy looking for ever-scarce government jobs, while the Internet gives them some opportunities the government cannot provide. So, the African governments should set up ad hoc committees and organisations that will re-orientate its youths from the rural areas to the cities, by educating them on the opportunities available in the 4IR.

African countries can prepare their youths for the 4IR by making them part of their 4IR Research and Development (R&D). The truth is that many youths have creative and innovative capacities to develop technologies that can help Africa seize the opportunities provided by the 4IR stage. We continue to hear numerous news of mobile apps, software, and 4IR technologies developed by African youths. For instance, in 2018, Catherine Ngirwa carried out research on Pesticides Authenticity Verification under the supervision of Dr Mussa Ally Dida. Both Tropical Pesticides Research Institute (TPRI) and Tanzania Telecommunications Corporation Ltd (TTCL) adopted her research work (Ngirwa & Dida, 2018).

Besides, the current gerontocratic leadership in Africa is due for a change. We can see that the majority of African leaders are elders. Nevertheless, it is time for the youths to contest, win, and hold political positions to impact the development of 4IR technologies in the continent positively.

However, the 4IR infrastructures cannot be provided by African governments alone. That is why partnership and collaboration are needed. Partnerships should be sought with national and multinational companies located in respective countries. Companies have always provided numerous social and infrastructural technologies in their host countries. For instance, in 2015, IBM initiated a programme called "IBM Digital – Nation Africa." It cost IBM $70 million, with the plans to empower up to 25 million African youths with digital, cloud, and cognitive Information and Technology (IT) skills (IBM Corporation, 2017).

References

African Development Bank Group. (2020, February 3). *Provide young people with new skills to meet the needs of a 21st century labor market, African Development Bank report urges.* African Development Bank. Retrieved May 13, 2022, from https://www.afdb.org/en/news-and-events/press-releases/provide-young-people-new-skills-meet-needs-21st-century-labor-market-african-development-bank-report-urges-33910

Annabel, S., Isabelle Carboni, C., & Caribou, D. (2020, November 23). *How to Develop Work-Ready Digital Skills in African Youth*? YouthPower. Retrieved May 13, 2022, from https://www.youthpower.org/resources/how-develop-work-ready-digital-skills-african-youth

Canagarajah, S. (2014, December). *Realizing Africa's youth potential: Africa needs investors to create jobs for its youth and develop skills*. World Bank Blogs. Retrieved June 23, 2022, from https://blogs.worldbank.org/developmenttalk/realizing-africas-youth-potential-africa-needs-investors-create-jobs-its-youth-and-develop-skills

Hastings-Spaine, N. (2021, May 21). *Leading South African Youth into the 4th Industrial Revolution*. Built In Africa. Retrieved June 23, 2022, from https://www.builtinafrica.io/blog-post/diketso-setho-ghetto-heroes-tech-innovation

IBM Corporation. (2017, February 8). *IBM Launches Digital - Nation Africa: Invests $70 million to bring digital skills to Africa with free, Watson-powered skills platform for 25 million people*. PR Newswire. Retrieved June 23, 2022, from https://www.prnewswire.com/news-releases/ibm-launches-digital---nation-africa-invests-70-million-to-bring-digital-skills-to-africa-with-free-watson-powered-skills-platform-for-25-million-people-300403817.html

Kalan, J. (2013, April 11). *African youth hungry for connectivity*. Africa Renewal. https://www.un.org/africarenewal/magazine/december-2016-march-2017/african-youth-hungry-connectivity

Kariba, F. (2020, July 30). *The burgeoning Africa youth population: Potential or challenge?* Cities Alliance. Retrieved May 13, 2022, from https://www.citiesalliance.org/newsroom/news/cities-alliance-news/%C2%A0burgeoning-africa-youth-population-potential-or-challenge%C2%A0

Karuitha, J. (2020, July 31). *Youth and the digital economy in Africa*. Africa Portal. Retrieved May 13, 2022, from https://www.africaportal.org/features/youth-and-digital-economy-africa/

Konrad, A. (2016, June 16). *Andela raises $24 million from Mark Zuckerberg and Priscilla Chan's fund to train African engineers*. Forbes; Forbes. https://www.forbes.com/sites/alexkonrad/2016/06/16/andela-raises-24-million-from-zuckerberg-chan-initiative-and-gv/?sh=2debb35811e0

MINDS Connect. (2018, August 4). Africa's challenges and opportunities in technology and infrastructure. MINDS – Mandela Institute for Development Studies. Retrieved May 13, 2022, from https://minds-africa.org/africas-challenges-and-opportunities-technology-and-infrastructure/

National Youth Policy of Ghana. (2010). Towards an Empowered Youth, Impacting Positively on National Development. Ministry of Youth and Sports Development, Ghana. Retrieved May 13, 2022, from https://www.youthpolicy.org/national/Ghana_2010_National_Youth_Policy.pdf

National Youth Policy of Nigeria. (2019). Enhancing youth development and participation in the context of sustainable development. Federal Ministry of Youth and Sports Development, Nigeria. Retrieved May 13, 2022, from https://www.prb.org/wp-content/uploads/2020/06/Nigeria-National-Youth-Policy-2019-2023.pdf

Ngirwa, C. & Dida, M. (2018). An ICT Based Solution for Pesticides Authenticity Verification: A Case of Tanzania. *Journal of Information Systems Engineering & Management*, 3(4), 27-35. Retrieved June 23, 2022, from https://www.researchgate.net/publication/328866577_An_ICT_Based_Solution_for_Pesticides_Authenticity_Verification_A_Case_of_Tanzania

Nwokoma, C. (2021, July 15). *How technology is helping African youths gain skills for employment and entrepreneurship*. Techpoint Africa. https://techpoint.africa/2021/07/15/technology-employment-entrepreneurship/

Population Reference Bureau. (2019). *Africa's future: Youth and the data defining their lives*. PRB. Retrieved May 13, 2022, from https://www.prb.org/resources/africas-future-youth-and-the-data-defining-their-lives/

South African Government. (2020, July 22). *Minister Angie Motshekga: Basic Education Department Budget Vote 2020/21*. South African Government.

Retrieved June 23, 2022, from https://www.gov.za/speeches/minister-angie-motshekga-basic-education-dept-budget-vote-202021-22-jul-2020-0000

United Nations. (2015). Youth population trends and sustainable development. *Population Facts, POPFACTS, No. 2015/1*. Retrieved June 23, 2022, from https://www.un.org/esa/socdev/documents/youth/fact-sheets/YouthPOP.pdf

United Nations. (n.d.). *Youth*. United Nations. Retrieved June 23, 2022, from https://www.un.org/en/global-issues/youth

Chapter 21

IT and Universities in Africa

As I have said repeatedly in this book, technology has been instrumental to the world's development and in the economic progress of nations. All sectors of the economy, including finance, energy, healthcare, manufacturing, transportation, real estate, entertainment, communication, and others, have all been transformed due to technological inventions and discoveries through the ages.

In like manner, the wealthiest nations have attained their status because of their ability to exploit technology and harness it to power the various sectors of their respective economies. They invest in developing technology infrastructure and research and development, broadening the scope of innovation that drives economic growth. Given their investment in technology research, they were well-positioned to take advantage of the First, Second and Third Industrial Revolutions to drive economic progress.

The emergence of the 4IR has ushered in an era of astonishing technological wizardry. Stunning technological leaps are being achieved due to progress in artificial intelligence, the Internet of Things, blockchain technology, big data, robotics, 3D printing, driverless vehicles and others. Already, homes are being remote-controlled, allowing people to regulate heat and check security around their homes while they are away. There has been much progress in developing autonomous vehicles, while artificial intelligence and significant data breakthroughs are reshaping how we treat diseases.

However, while Western countries have consistently advanced in technology and innovation, African countries have continued to lag, even falling further down when measured against the indices for determining the level of technological advancement of nations. A significant reason for Africa's lack of technological innovations is the lack of skilled manpower. Skilled human capital is imperative to stimulate the development of science and technology for African countries, to leapfrog economic growth and achieve equal ranking with the Western world.

Although Africa has significantly benefitted from the uptake, diffusion and assimilation of technology across various sectors of the economy, its countries need to move beyond merely consuming technology products to becoming creators, just like the developed countries. The critical step towards this transition is the development of skilled manpower to drive technology and innovation on the continent.

The case of "newly industrialised countries" (NICs) such as Hong Kong, Singapore, South Korea, Taiwan, Brazil, China, India, Malaysia, the Philippines, Thailand, and South Africa serves as excellent examples of how African countries can move up the technology ladder. Starting in the 1970s and 1980s and later in the 2000s, the NICs listed above leveraged technology from Western countries to stimulate the acquisition of technical skills and grow their industrial base, after which they created technological breakthroughs (CFI Team, 2021).

There were concerted efforts to create a large base of educated citizens that would give rise to a science-literate and technically-skilled workforce. Enrolment of children in school was encouraged, and university graduates were encouraged to pursue further studies, including PhDs, to undertake research in various areas of science and technology that could potentially lead to discoveries and innovations.

Creating a Large Base of Science-Literate and Technically-Skilled Citizens in Africa

Like NICs of the 1970 – 1980s and 2000s, African countries must build a solid and virile scientific base to anchor national and continental efforts to boost technology and innovation to leapfrog economic development. As the experience of developed countries and NICs has shown, no nation can achieve technological advancement without sound and vibrant scientific and technical education.

Developed countries and NICs gave pre-eminent attention to science in their school curriculum to build a solid scientific and technological education base. The government made it obligatory for every child to attend school. Starting at an early age, students were exposed to scientific concepts to deepen their knowledge and stimulate their interest in science and technology.

This is at variance with the situation in Africa, where it was on record in 2019 that "over one-fifth of children ranging in age from around 6 and 11 are not in school, followed by one-third of young people aged between 12 and 14," and that "nearly 60% of young people aged between about 15 and 17 are out of school" (Welle, 2022).

Besides primary education, African countries must increase enrolment in secondary schools, mainly because the higher the number of young people who enrol in and complete secondary school, the higher the chances of such young people proceeding to acquire tertiary education in science and technology.

Tertiary education is equally important. In particular, African universities have a huge role to play in stimulating the technological development of the continent. They can churn out the skilled workforce that is highly needed for mastery of technology and economic growth. At the moment, it is evident that African countries lack a large base of science-literate citizens and the skilled workforce required to pursue a programme of technological advancement.

However, with over 1,800 universities across the continent, Africa has enough tertiary institutions to produce a sufficient number of scientists and technology professionals who will become creators of innovative technology products and, thus, leapfrog the continent from a mere consumer of technology to a producer (Kamer, 2022).

How African Universities Can Drive Technology and Innovation

Universities are instrumental in fostering knowledge acquisition and skill development. A highly skilled workforce with high literacy in science and technology is a sine qua non for innovation and economic growth. African universities have been responsible for the supply of high-calibre human capital on the continent for upward of 50 years.

In the 1960s and 1970s, African universities had a reputation as centres of excellence with a standard that compared with the best around the world. The University of Ibadan, the University of Makerere, Fourah Bay College and the Université de Dakar are a few of the institutions where teaching and research were at a standard that matched the best schools in the world.

By the 1980s, the situation with African universities changed when an economic crisis hit most African countries owing to the implementation of structural adjustment policies suggested by the IMF. Consequently, priority was given to primary and secondary education. Resources allocated to universities dwindled, resulting in a plunge in the quality of post-secondary education. Ultimately, many universities stopped undertaking research that could have led to innovation as there were no sufficient resources to fund such research endeavours.

Nevertheless, universities remain the best hope for African countries to provide their citizens with the requisite skills and knowledge to engage in research and innovation. As a continent that aspires toward technological development, universities have an essential role in developing relevant and competitive human capital in the Fourth Industrial Revolution era. Universities must remain centres of excellence for knowledge production and nurture and sustain research and innovation to support Africa's quest to leapfrog technologically.

It is important to note that the academic community, the government, and the technology industry, form the triumvirate that constitutes the world's innovation triangle. If universities cease to function optimally as centres of excellence for research and innovation, the tripod would not be able to stand; there would be an imbalance in the technology and innovation ecosystem.

Around the world, many universities are setting the pace for innovation, creating high-tech products that further advance humanity. Even though university enrolment has consistently increased over the years, research and innovation are still in their infancy, particularly in Sub-Saharan Africa. Universities in the region are constrained by inadequate funding, limiting their capacity to fulfil their objectives and mission. It will be a significant step forward for

Africa if the impediments that restrict universities on the continent are removed so that each institution will leverage its intellectual capacity to drive research and innovation, as is the case with universities in the West.

One such university is the Technion-Israel Institute of Technology which has contributed significantly to the technological advancement of Israel and the world. The innovations spearheaded by the institution are a testament to the importance of universities in driving innovation and stimulating technological development.

To highlight the contributions of Technion-Israel Institute of Technology to Israel's technological advancement, as of 2012, the institution has graduated 67,000 software developers and engineers who are making significant contributions and adding great value to humanity (Frenkel & Maital, 2012). Additionally, more than 90,000 Technion graduates set the stage for Israel's global innovativeness, primary research, and rapid industrialisation.

Specifically, research in the department of aeronautical engineering at Technion has led to critical breakthroughs that have been of great help in defending Israel against attacks from hostile neighbouring countries. Meanwhile, Israel's electron and software industry which generates $20 billion annually is populated by talented Technion graduates. Likewise, 12% of Technion graduates work in R&D, while out of all Technion graduates employed in industry, 75% are used in the high-tech sectors (Frenkel & Maital, 2012).

There Should Be Concerted Efforts by Government to Fund Research

For African universities to make an impact similar to Technion-Israel, their respective governments must make concerted efforts to fund research and development activities. Technology-oriented research facilitates innovation and lays the groundwork for systemic technological transformation and economic development. Since Africa needs to leapfrog economic growth, it is only logical to expect that African countries will appreciate the need to devote resources to deepen technology research capabilities on the continent.

When the statistics are considered, the need for Africa to aggressively drive technology research becomes urgently imminent.

In 2017, Africa was 15% of the world's population while having 25% of the global disease. However, the continent contributed only 2% of the global research output. Much of the medical research on the continent does not prioritise the prevalent diseases (Kariuki & Kay, 2017).

As of 2018, Africa had 198 researchers per one million people compared with approximately 900 in Brazil, 5,331 in Japan, 6,803 in Singapore, 4,603 in the UK and 4,412 in the USA (World Bank Open Data, 2022).

The dismal presence of engineers and scientists in Africa stems from a pervasive lack of investment in science and technology. The lack of investment in technology research has stifled Africa's economic transformation by making it impossible to build high-productivity sectors that would have created wealth within the economies of respective African countries. Also, there is an absence of the technological infrastructure necessary for innovation. Therefore, the continent has continued to rely on the out-modelled practice of extracting and exporting raw resources and importing finished products, which essentially amounts to exporting wealth and importing poverty.

To arrest the debilitating descent into poverty, African states must commit a decent percentage of their income to funding research and development to build the necessary infrastructure to fuel innovation. As of 2020, Africa was responsible for only 0.9% of total global R&D, despite having more than 20% of the world's population. Conversely, 25%, 23% and 7.8% of global R&D spent are attributed to the USA, China and Japan, respectively (Heney, 2020).

African countries must contribute more of their real GDP to research to improve the continent's involvement in global scientific and technology research. As of 2019, African countries spent only 0.42% of their GDP on research, a significant shortfall from the global average of 1.7% (Kyobutungi, 2021). Also, the government should provide incentives to research by committing resources to research endeavours by corporate organisations, which could be commercially exploited.

Furthermore, African states should enlarge the pool of researchers on the continent by providing scholarships to citizens with the intellectual capacity to undertake PhDs in science and technology and should commit funds to increase the capacity of existing research institutions to undertake quality research relevant to the challenges confronting the continent, to generate innovative solutions that will be useful to Africa and the rest of the world.

With the government of various African countries committing more resources to funding research institutions and providing an enabling environment, it is not inconceivable that many Diaspora scientists and technology professionals will elect to return to undertake research in Africa. Generally, the best-trained and most talented researchers tend to move to environments where there is the existence of modern equipment and sufficient funding – and perhaps, more importantly, where there is a large pool of fellow researchers whose knowledge, skills and experience they can leverage in the course of their work.

Collaboration between University Research and Industry

Another way to boost technological development in Africa is via university-industry collaboration. Universities serve as centres for generating new knowledge through research, while the industry focuses on applying the ability to create products that have commercial value. As such, a collaboration between industry and universities can accelerate advances in technology and innovation.

Historians have established that universities and industries have collaborated for more than 150 years. University researchers and their counterparts in European companies started collaborating as far back as the 1800s. The university-industry collaboration began during the technological revolution in the United States of America. Currently, the focus areas for most university-industry collaboration in Western countries are science and technology, especially medicine, information technology, agriculture and engineering.

Like their Western counterparts, African universities can collaborate with companies on the continent on research that will be developed into practical applications that will boost technology and innovation. If properly harnessed, university-industry collaboration could become a strong force that will drive the efforts of African countries to leapfrog technological development.

Both academia and industry have much to gain from collaboration. For academia, the funds made available by the sponsoring companies make it possible for universities to acquire cutting-edge equipment for their laboratories. The availability of innovative equipment allows university scientists to undertake research into challenging issues and generate results that have real-world applications that could be commercialised.

For a sponsoring company, partnering with academia gives them access to current research that might spur the development of innovative processes or products. By accessing recent academic studies, companies drastically reduce the cost of R&D and shorten the time it takes to move a potential outcome from the R&D laboratory to the market. Developing innovative techniques and products and reducing the cost of R&D gives a business a competitive advantage and significantly improves the bottom line.

How to Promote University-Industry Collaboration in Africa

There is great potential for universities and research institutions in Africa to contribute significantly to the continent's growth of science and technology. This potential was demonstrated during the COVID-19 pandemic when "researchers at the DiaTropix facility of the Institut Pasteur de Dakar leveraged their AIDS and Ebola experience to develop an immune-based $1 diagnostic kit that can test for COVID-19 in 10 minutes" (Peeling & Sall, 2020).

Besides the feat achieved at the DiaTropix facility of the Institut Pasteur de Dakar, "engineering students at the Ecole Superiéure Polytechnique in Dakar have also built a multifunctional medical robot to lessen the workload of healthcare workers" (Peeling & Sall, 2020). Following collaboration between government research institutes and some of the country's universities, Nigeria became the first in Africa to sequence the SARS-CoV-2 genome.

These successes point to the possibility of African universities spearheading headline-making innovations that could solve some of the most challenging problems and ultimately ignite a technological revolution on the continent. However, it will take collaborative synergy between universities and industry to scale innovations resulting from research by academia. At the moment, such collaborations are non-existent.

The following measures can be adopted to foster collaboration between African universities and industry.

1. **Setting Clear Objectives:** From inception, universities and industries should clearly define the objectives of their collaboration. The objectives should be apparent if the collaboration is for a short-term consulting engagement. And if it is an R&D undertaking that would take considerable time, both parties should agree on the objective of the project they are embarking on and the duration it would take before they start to see any results.
2. **Re-Positioning of Interface and Partnership Offices:** Several universities across Africa have an interface office for consulting, technology transfer, intellectual property licensing and coordination of research collaboration with industry. These offices are usually headed by academics and supported by an administrative staff recruited by the university. Consequently, interface offices focus more on driving the agenda of universities than on operating as partnership offices that incorporate industry objectives for collaborating with academia. Universities in Africa should re-position their interface offices to serve as centres that align university and industry interests for a successful industry-academia partnership.

   In addition to seeking to harmonise university and industry interests, interface offices should serve as business development units, constantly courting potential industry partners for mutually beneficial collaborations.
3. **Consider the Possibility of Global Partnerships:** Just like scientists and engineers of African origin involved in research and development are spread all over the world, African universities should consider the possibility of building networks that would provide opportunities to partner with corporate organisations beyond the shores of Africa.

   The partnership between Mologic and the DiaTropix department, of the Institut Pasteur de Dakar, for developing COVID-19 test kits, is good evidence of the opportunities for African universities to seek partnerships with industry on a global scale.
4. **There Should Be Provision for Short-Term and Long-Term Collaboration:** One of the reasons for the lack of collaboration

between universities and industry in Africa is that universities usually focus on R&D with a long-term perspective. In contrast, industries focus on short-term projects that will bear fruits within a specified period. Industries have repeatedly accused universities of allowing bureaucratic processes to cause delays to the pace of R&D, thus delaying the achievement of their objectives.

To resolve this situation, universities should develop research policies that support short-term, industry-friendly partnerships and procedures to facilitate long-term collaborative relationships that cater to non-disclosure agreements, intellectual property ownership and patent licensing. A long-term relationship model will not require negotiation when another research project is undertaken.

The Need for African Universities to Update Their IT Curriculum

As I have explained throughout this book, we have entered the era of the Fourth Industrial Revolution. Mainstream technologies of the past are fast becoming intelligent-based. Artificial intelligence, the Internet of Things, blockchain technology, big data, quantum computing, robotics, autonomous cars, and the exponential technologies which underpin the Fourth Industrial revolution are gradually becoming ubiquitous. Over the next ten years, artificial intelligence will penetrate virtually every industry and become as ever-present in every industrial process as electricity.

There is no gainsaying that the university and college IT curriculum in Africa is outdated, so it is not fit for purpose in the era of the Fourth Industrial Revolution. The curriculum is filled with concepts that have become obsolete and are irrelevant to the needs of the Fourth Industrial Revolution. For most African universities, the IT curriculum used for training students does not match the needs of the current digitalised labour market. As such, African education policymakers must rejig the IT curriculum of universities on the continent to make it more attuned and responsive to the emerging technologies of the Fourth Industrial Revolution.

Such a re-organisation of the curriculum is needed to prepare African students to function optimally in a digitalised workplace, even going as far as undertaking practical research which could lead to innovative discoveries that would advance the frontiers of science and technology. The governments of African countries should ensure that substantial changes are made to the IT curriculum of universities to enable students to develop capacity in the emerging areas of AI, the internet of things, big data, blockchain, nanotechnology, genomics, and others.

The workplace of the Fourth Industrial Revolution requires workers who are competent in using and developing various technologies. This makes it imperative for the IT curriculum of African universities to reflect these expectations in preparation for what lies ahead for the youth of Africa. In the years ahead, the African industrial sector will only have a pool of tech-savvy graduates to recruit if the IT curriculum is adapted to be responsive to the demands of the Fourth Industrial Revolution.

In striving to create a responsive IT curriculum, policymakers should ensure that they adapt it to address the needs of students and the circumstances of African society. The curriculum should be such that it will equip students with the skills to solve both current and anticipated organisational problems. The sections of the current curriculum that are not useful in the Fourth Industrial Revolution should be excised to create room for more relevant topics. While the curriculum should be adapted to suit the unique needs of Africa, the content should match the highest standard anywhere else in the world. It must be continuously updated to include new thoughts and knowledge in IT. This is to ensure that the curriculum will never become obsolete again.

In this book, I have captured in one volume all that stakeholders in the African technology ecosystem need in order to play in this era of the 4IR. The book starts by looking at African antiquities for achievements that will motivate today's Africans. Then, it provides a good picture of all the information technologies. It presents the salient goals and actions that need to be undertaken by individuals, organisations, countries and Africa as a continent for it to be part of the train of this new industrial revolution. If the continent, governments, organisations and individuals take this book's message seriously and recognise the urgency required for implementing all that is being advocated, Africa will indeed ride the wave of the 4IR.

References

CFI Team. (2021, February 15). *Newly Industrialized Country (NIC) - Overview and Characteristics*. Corporate Finance Institute. https://corporatefinanceinstitute.com/resources/knowledge/economics/newly-industrialized-country-nic/

Frenkel, A., & Maital, S. (2012, March 2). *Technion's Contribution to the Israeli Economy through its Graduates. Neaman Institute Working Paper.* Technion - Israel Institute of Technology. https://www.technion.ac.il/en/2012/03/hot-technion-facts-on-startups-innovation-israel-science-technology/

Heney, P. (2020, March 19). *Global R&D investments unabated in spending growth.* Research & Development World. https://www.rdworldonline.com/global-rd-investments-unabated-in-spending-growth/

Kamer, L. (2022, February 1). *Number of universities in Africa by country*. Statista. https://www.statista.com/statistics/1242428/number-of-universities-in-africa/

Kariuki, T., & Kay, S. (2017, May 3). *There are not enough scientists in Africa. How can we turn this around?* World Economic Forum. https://www.weforum.org/agenda/2017/05/scientists-are-the-key-to-africas-future/

Kyobutungi, C. (2021, May 18). *African countries must muscle up their support and fill massive R&D gap*. The Conversation. https://theconversation.com/african-countries-must-muscle-up-their-support-and-fill-massive-randd-gap-161024

Peeling, R. & Sall, A. (2020). *Opportunity in Every Difficulty: Lessons for Africa to take from the COVID-19 pandemic*. Lab Culture, African Society Laboratory Medicine, Issue 23. https://aslm.org/wp-content/uploads/2020/10/Lab-Culture-Issue-23-V2.pdf?x55504

Welle, D. (2022, January 24). *Why the right to education remains a challenge in Africa.* Deutsche Welle (www.dw.com). https://www.dw.com/en/africa-right-to-education-remains-a-challenge/a-60518000

World Bank Open Data. (2022, June). *Researchers in R&D (per million people).* World Bank Open Data. https://data.worldbank.org/indicator/SP.POP.SCIE.RD.P6

ACKNOWLEDGEMENTS

A book of this magnitude, scope, and depth requires many individuals' direct and indirect assistance and contributions. It would thus be impossible to name all those who contributed to this work. However, I will venture to name a few here.

For a long time now, I have been convinced that African countries need to experience accelerated but sustainable development that would put them in the same bracket as other parts of the world. To do so, I am convinced that technology needs to be adequately harnessed and utilised as a catalyst.

One African who encouraged me to be so convinced was the late Prof Calestous Juma through his writings and discussions with me, especially when he guided me on my first information technology (IT) book in Africa. He immensely motivated me, so the idea of leveraging technology as a catalyst for economic development became ingrained in me. Thus, it has shaped my worldview and impacted how I think about the African renaissance and reawakening. I thank Prof Juma for encouraging me in this direction.

I think highly of Professor Juma. He was an intellectual giant and knowledge source on a technology-led development solution for the continent. When we were planning the inaugural Digital Africa Conference in 2013, he and I exchanged emails about the conference's objectives and what I hoped to achieve. He highlighted how long I had been passionate about a technologically developed Africa, referencing my book, *Information Technology and Africa: Challenges and Opportunities*, published in 1993 by ACTS Press, Nairobi, Kenya.

He wrote, "Dear Evans, first let me say that you were way ahead of the time. So far out that only a fool like me would have agreed to publish your book. I would like to attend your conference, but unfortunately, the dates do not work.... There are so many areas where Africa was always ahead, but we never had high-level champions.

Your book is an example."

I received tremendous encouragement and stimulation for this book through the shared knowledge and questions raised during the nine Digital Africa conferences and exhibitions in Abuja and Lagos since 2013. I would like to thank my colleagues who assisted in planning and executing each of the Digital Africa Conference events.

Therefore, I graciously salute my Digital Africa colleagues and core friends, namely: Prof Isa Ali Ibrahim Pantami, Prof Umar Garba Danbatta, Chief Ernest Ndukwe, Sir Demola Aladekomo, Inuwa Kashifu Abdullahi, Prof Pat Utomi, Robin Raskin, Dr Mayuri Odedra, Biodun Omoniyi, Adekunle Ajiboye, Senator (Dr) Datti Baba-Ahmed, Andy Chen, Stephen Ibaraki, Fred Bussemaker, Dr Armstrong Takang, Bosun Ayeni, Nyema Weli, Liz Donaholu, Late Don Aholu, Reginald Hanotu, Uzo Ndidigwe, Eddie Okon, Nneoma Ofodile, Shade Ade Joseph, Chris Uwaje, Dumebi Obodo, Pius Okigbo Jnr, Chinenye Uzoukwu, Yele Okeremi, and Bayero Agabi.

I would also like to thank all the Sponsors and Partners of the annual Digital Africa Conference and Exhibition, a few of which are Baze University, VDT, First Bank, Chams Plc, Chams Access, Zenith Bank, NITDA, MTN, Compumetrics Solutions Ltd and Unity Bank.

This book required the elicitation of an enormous amount of data and information. It, thus, needed a precise and copious amount of research work. I was fortunate to have, and I can never, therefore, thank them enough, a brilliant research team headed by Tervel T. Kejih and Dr Kayode Odeyemi. They practically became my mental prosthetics, and so did most of the heavy lifting of the research work for this project. They scouted for the supporting data and dug very deep to unearth current developments in all the domains of discourse for this book.

I specifically thank the National Information Technology Development Agency (NITDA) of Nigeria, which has been doing an incredible job of managing the Information Technology industry in Nigeria towards building a mature Digital Economy for the country, and for its yearly assistance to Digital Africa in organising and hosting the annual digital event. I am also grateful to the National Communications Commission (NCC) of Nigeria for the excellent job over the years in turning around the Nigerian Telecommunications industry, which is a vital component of the Nigerian Information Technology industry or ecosystem.

I want to use this opportunity to most graciously thank Mallam Mohammed Hayatudeen. He believed in me in terms of what I could do in technology management for economic development, thus encouraging me so much in that direction. He went to South Africa to persuade me to return to Nigeria. He remains one of the best Nigerian turnaround CEOs. He will go to any length to get whomever he thinks can assist him in executing whatever goals his organisation needs to achieve, without considering their place of origin or religious inclination - or lack of it. Similarly, Mr Jim Ovia and Mallam Falalu Bello also believed in me. Accordingly, they encouraged me to accomplish legacies that will inspire the youths of Nigeria and Africa to develop moon-shot mindsets in the use of technology to achieve economic development for the countries of Africa; hence, this book.

Finally, but very importantly, I want to thank my wife, Margaret, for always motivating me to put down some of my thoughts on paper. Also, she is always available to let me know when I am on the right path and when I need to rethink. I also appreciate my lovely children, Chizi and Chioma, for continuing to be part of the inspiration that fires my desire and search for a new Africa.

OTHER BOOKS BY THE AUTHOR

1. Woherem, E. E. (1993): *Information Technology in Africa: Challenges and Opportunities*. ACTS Press, Nairobi, Kenya, and RIKS, Maastricht, Netherlands.
2. Woherem, E. E. (Ed) (2000): *Information Technology in the Nigerian Banking Industry*. Spectrum Books, Ibadan, Nigeria.
3. Woherem, E. E. and Odedra, M (Eds) (2016): *Expanding Africa's Capacity for Digital Technology, Proceedings of the 2015 Digital Africa Conference and Exhibition*. Nsemia Inc Publishers, Nairobi, Kenya.

ABOUT THE AUTHOR

Dr Evans Woherem is an award-winning African Technology researcher, analyst, and writer. He has authored or co-authored three books and numerous academic papers on various areas, including Artificial Intelligence, Expert Systems, Software Engineering, project management, IT Strategy Planning and Technology in Banking. He worked for Lloyd's Register (UK) and ABSA bank (RSA) at technical and management levels.

He was a highly successful Director of Technology at First Bank (Nigeria) and Unity Bank Plc (Nigeria). He was the pioneer Chairman of Interswitch Nigeria Ltd, which is today a Unicorn. He was also the Chairman of Chams Access Ltd. He is currently the Chairman of Digital Africa, Compumetrics Solutions Ltd and Northlink Insurance Brokers Ltd. He holds a PhD on Impact of Expert Systems. He is a Fellow of the Nigerian Computer Society, married with two children, and an alumnus of Harvard Business School.

INDEX

www.ingramcontent.com/pod-product-compliance
Lightning Source LLC
Chambersburg PA
CBHW031335230926
51848CB00002B/244
* 9 7 8 9 7 8 5 9 7 6 0 8 3 *